STRUCTURAL SHEAR JOINTS

ANALYSES, PROPERTIES AND
DESIGN FOR REPEAT LOADING

George T. Hahn

Mechanical Engineering, Vanderbilt University

Kaushik A. Iyer

U.S. Army Research Laboratory

Carol A. Rubin

Mechanical Engineering, Vanderbilt University

WOODHEAD PUBLISHING LIMITED

Library of Congress Cataloging-in-Publication Data

Hahn, George T.
  Structural shear joints : analyses, properties and design for repeat loading / George T. Hahn, Kaushik Iyer, Carol A. Rubin.
     p. cm.
  Includes bibliographical references and index.
  ISBN 0-7918-0238-8 (hardcover)
  1. Shear (Mechanics) 2. Bolted joints. I. Iyer, Kaushik. II. Rubin, Carol A. III. Title.

  TA417.7.S5H34 2005
  621.8'82—dc22   2005016167

British Library Cataloguing in Publication Data
A catalogue record for this book is available from the British Library.

Woodhead Publishing ISBN-13: 978-1-84569-119-6
Woodhead Publishing ISBN-10: 1-84569-119-9

Cover graphic: Computed maximum principal stress contours in a butt joint with 0.6% interference that is subjected to a remote tensile stress of 80 MPa

TABLE OF CONTENTS

PREFACE

Engineers employ shear connections — riveted and bolted butt and lap joints — in a wide range of structures and machines. Ordinary stress analyses of joints are unable to define the features that ultimately govern fatigue and fretting and provide bases for design. Detailed analyses have only become possible in the past decade with advances in finite element methods and computing capabilities. This text places into context the results of over 150 detailed, 2D and 3D finite element analyses of aluminum and steel shear connections. The text examines the following joints, both single and multiple row, fasteners and modes of load transmission:

butt joints	bearing mode	pins
lap joints	clamped mode	standard head fasteners
attachment joints	adhesives mode	countersunk fasteners
and doublers	bearing mode with hole expansion	self piercing rivets
		interference fasteners

The analyses examine the fastener joint geometry, the adjacent region of the panel and the important design variables. The text connects global features such as nominal stress, excess compliance and fastener load, with the local conditions that affect fatigue and fretting. Among the important local conditions are the contact pressures, interface slips, residual stresses and the intensity and location of the stress concentrations. The text is divided into 3 sections designed for different audiences:

Section 1

Engineers and technicians seeking a better understanding of shear connections. The first section introduces the different joints and identifies the important variables. It offers stress contours and distorted meshes that complement easily grasped descriptions of the behavior of panels and fasteners for the different modes of load transmission.

Engineers preparing to design shear connections resistant to fatigue. The pros and cons of the butt and lap joints for the different load transfer modes are discussed and a fail safe strategy is identified in the first section. In addition, a simple and general FS-SCF (Fatigue Strength-Stress Concentration Factor) analysis useful for preliminary design is identified and validated. The analysis offers estimates of either the fatigue strengths of joints or the joint geometry meeting a given fatigue strength requirement. Expressions for estimating the bolt tension of clamped joints and the fastener load and SCF of multiple-row joints are derived in appendices.

Engineers and technicians interested in fretting and in the fatigue strength of shear connections. An extensive selection of fatigue strength measurements for riveted, bolted, and bonded aluminum and steel joints is offered in the first section. The measurements are compared with the forecasts of the FS-SCF estimation procedure. The contributions of fretting wear and fretting fatigue are also examined.

NDE practitioners interested in knowing where to look for fatigue cracks and fretting damage. The stress concentrations in shear joints are responsible for fatigue damage and crack initiation. The locations of the stress concentrations, which depend on the joint type and mode of load transfer and are affected by hole expansion, are identified in the first section.

Section 2

Engineers and researchers wishing to access more finite element analysis findings. A handbook-type, more detailed summary of the results of finite element calculations for fasteners and panels is presented in Section 2 of the book.

Section 3

Engineers and researchers seeking to perform finite element analyses of shear connections. To assist workers with computations, the finite element models, meshes, boundary conditions and procedures used by the authors are described in Section 3. This includes TALA, a Thin Adhesive Layer Analysis for modeling joints fastened with either adhesive alone, or with rivets (or bolts) combined with a sealant or adhesive. Material models and parameters employed in the calculations are listed. Validation of the finite element methods in the form of comparisons with closed form analyses, other computations and measurements is presented.

ACKNOWLEDGEMENTS

The authors are indebted to John Wikswo of the Vanderbilt University Physics Department who is responsible for inspiring the initial phases of the research endeavor described in this book. The authors acknowledge the important contributions of the graduate students who contributed to this research. In alphabetical order they are: former PhD students Charoenyut Dechwayukul, Thongchai Fongsamootr, and Notsanop Kamnerdtong; and former MS students Khalid Al-Dakkan, Felicia Brittman, Yun Huang, and Charat Loha. They also wish to thank Pedro Bastias, a former PhD student and colleague, who provided much help and advice for the graduate students.

Computations were performed with the ABAQUS finite element program provided by Abaqus, Inc. Most computations were performed on parallel systems in 3 major centers: the National Center for Supercomputing Applications in Illinois (SGI Power Challenge Array), Center for Advanced Computing at the University of Michigan in Ann Arbor (IBM SP) and Department of Defense Major Shared Resource Center at the U.S. Army Research Laboratory in Maryland (IBM SP and SGI Origin).

Riveted lap joint specimens with and without a sealant were prepared Textron Aerostructures, Inc., Nashville, TN. with the help of Tom Warrion, Mike Floyd and Paul Nelson. Bill Keller of PRC-DeSoto International, Inc. (formerly Courtalds Aerospace) provided the test sealants. Bill King of Emhart Teknologies, Mt. Clemens, MI and Daniel Hayden of General Motors Corp., Warren, MI provided assistance with fabrication and testing, respectively, of the Self-Piercing Riveted (SPR) joints.

Changchun Liu and Tuomas Wiste, graduate students at Vanderbilt University, and, Shirleen Johnson of the graphic arts division at the U.S. Army Research Laboratory provided significant assistance with the drawings in this book.

The work was supported by the Air Force Office of Scientific Research, Vanderbilt University, The Royal Thai Government and the General Motors Collaborative Research Lab (GMCRL) at the University of Michigan (Ann Arbor).

INTRODUCTION

This book draws on the results of 2- and 3-dimensional, finite element analyses of shear joints performed by the authors and graduate students between 1992 and 2001 [1-8]. Much of this work has been published [9-22], but many findings are reported for the first time. The preparation of this text prompted additional, and more detailed computations. The text presents these new findings which have modified some of the authors' earlier interpretations.

The analyses evaluate features relevant to joint performance with respect to fatigue and fretting failure. The computations treat the open-hole panel and seven different joint configurations, including eight ways of fastening them, and 15 other joint variables:

Joint Configuration
1. Panel with an open hole
2. Attachment joint
3. Butt joint
4. Single fastener row lap joint
5. Double fastener row lap joint
6. Adhesive lap joint
7. Double fastener row doubler

Fastener Type
H. No fastener (open hole)
P. Pin
S. Standard fastener head
C. Countersunk fastener head
A. Adhesive or sealant
SA. Standard fastener combined with adhesive
CA. Countersunk fastener combined with adhesive
SP. Self-Piercing rivet

Joint Variables

Panel material	Fastener material	Interface friction
Elastic or elastic plastic behavior	Fastener head shape	Uniaxial and biaxial loading
Panel width	Interference	Adhesive properties
Panel thickness	Fastener clamping	Adhesive thickness
	Number of fastener rows	

An alpha-numeric code employing the joint configuration number and fastener-type abbreviation (shown above) is defined in Appendix A. This code is used throughout the book to identify the calculational models employed.

The book is divided into 3 sections:

Section 1 (Chapters 1 – 5) describes the mechanical behavior of shear joints, the effects of major variables, failure modes, and design considerations for fatigue and fretting.

Section 2 (Chapters 6-12) offers handbook-like summaries of the results of the finite element analyses in the form of stress contour plots, graphs and tables, with the bulk of the findings presented in Tabular form in Chapter 12.

Section 3 (Chapters 13-18) Describes the computational methods and their validation. This includes the finite element models of joints and thin adhesive layers, material behavior modes, spring models of multiple row joints, and an analysis of the fastener clamping force.

SECTION 1

MECHANICAL BEHAVIOR OF SHEAR JOINTS AND DESIGN CONSIDERATIONS

BASIC FEATURES

1.1 ASPECTS OF SHEAR JOINTS

Shear joints are structural details for fastening panels together to form a load carrying connection. Such joints are also referred to as shear connections or splices and they can be fastened with rivets, bolts, adhesive or welds. The term *panel* is used here to mean either sheet, plate, beam, cylinder or vessel. The contrasting features of four common shear joints are illustrated in Figure 1.1:

Butt Joint. The butt or *double shear* joint, shown in Figure 1.1a, is formed by fastening two abutting panels with overlapping side panels. The external load is applied to the abutting panels, creating a state of nearly pure shear in two planes of the fastener shank and pure in-plane (2-dimensional) deformation in the two center panels. The bending of the fasteners and out-of-plane movements of the side panels are restricted by close-fitting fastener holes and presence of the middle panel, and are usually negligible.

Attachment Joint. This is a panel loaded along one edge and fastened to a relatively rigid supporting structure. The response is largely 2-dimensional. The limiting case, treated in this text is for an attachment joint whose fasteners and support structure are rigid and fixed in space.

Lap Joint. The lap or *single shear* joint is formed by fastening the overlapping edges of two panels. Figure 1.1c illustrates that the shear load applied to the fasteners by the panels tilts the fasteners and bends the panels out-of-plane. The fastener tilt angle can amount to 1°–3°; out-of-plane displacements of the panels to 1 mm to 2 mm. These are not negligible and are accounted for by the 3-dimensional analyses in this text.

Doubler. This is a panel fastened to a larger, main panel as a reinforcement. The connecting fasteners divert load to the doubler thereby reducing the stresses on the main panel. Although the magnitude of out-of-plane deformations may be small, the geometry is 3-dimensional and the analysis must treat it as such.

Multiple Row Joints. The shear connections in Figure 1.1 are depicted with a single row of fasteners for clarity. In practice, two or more rows of fasteners may be inserted to reduce

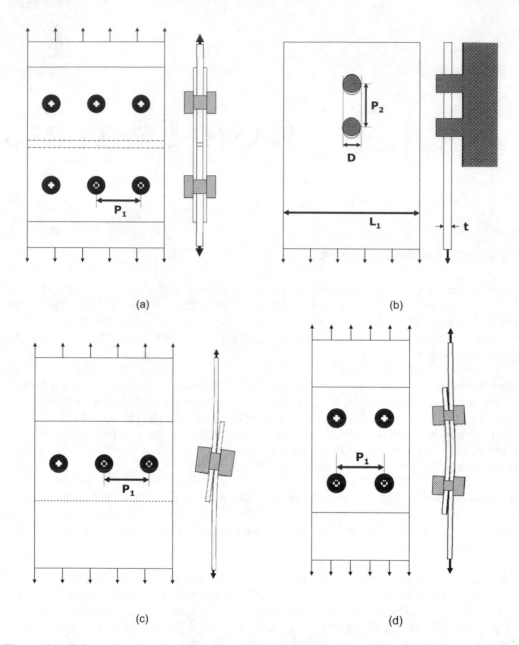

Figure 1.1. Schematic drawings of 4 shear joints: (a) butt joint, (b) attachment joint, (c) lap joint and (d) doubler.

the local stresses generated in the fasteners and the panels. Figure 1.2 illustrates a lap joint with three fastener-rows, and introduces the terminology: "leading" and "lagging", for the row of fastener *holes* (and not the fasteners themselves) closest to and furthest from the panel edge where the load is applied, respectively. The terms "front" panel and "back" panel

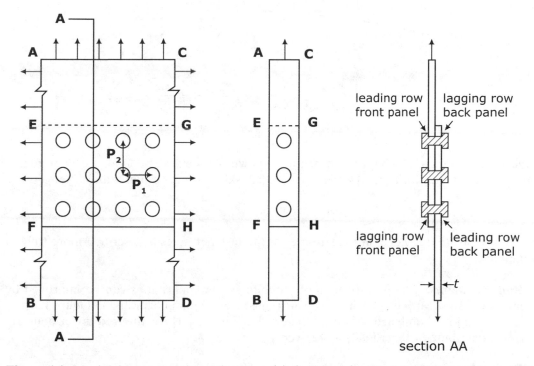

section AA

Figure 1.2. Model features and terminology: (a) (infinitely) wide joint. Edges AB and CD are stressed; edges EG and FH are stress free and (b) narrow joint with edges AB, CD, EG and FH stress free, and (c) Section AA with definitions of the leading and lagging row of holes of multiple row joints.

are used to identify the panels that can and cannot be visually inspected, respectively, when only one side of the joint is accessible.

An important feature of multiple-row joints is that the outer rows of fasteners, the two rows of fasteners connecting the leading and lagging row of holes, support the largest loads. At the same time, the local stresses generated in the panels adjacent to the leading row of holes are always higher than at the lagging holes. Analyses of the fastener loads and stresses in multiple row joints are examined in Chapters 2.6 and 17.

Wide and Narrow Joints and Biaxial Loading. In some applications, shear joints are narrow and formed by short rows of one or two fasteners. For others, like the cylindrical aircraft fuselage, the joints are wide with long rows containing scores of fasteners. This text draws on the two types of shear joint models illustrated in Figure 1.2 and described fully in Chapter 13. One is infinitely wide, has a repeat distance of P_1, and describes the conditions in the *interior* of a wide joint. The other is a model of a "narrow," single fastener joint of width, L_1, with stress free edges on all sides.

When wide joints are loaded in simple tension, the in-plane transverse contraction of the panels in the vicinity of the fastener holes is constrained and a state of biaxial stress is generated away from the free edges. The influence of this and other biaxial stress states and the

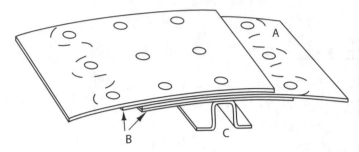

Figure 1.3. Example of a built-up shear joint. Shown is a section of a Douglas, longitudinal, lap-type skin-splice for an aircraft fuselage [23]: A-skin, B-finger doubler and C-longeron.

proximity of the free edge on the stressses at the fastener holes is described more fully in Chapter 2.7.

Built-Up Joints. The performance of joints for some applications can be improved by combining two shear joint elements. The airframe fuselage joint illustrated in Figure 1.3 consists of a lap joint combined with two "finger" doublers [23]. The contoured finger doublers serve to distribute fastener loads more evenly.

Design Aspects. Dimensions of the fastener relative to those of the panels, fastener geometry, fastener spacing, clamping and interference residual stresses due to fastener installation, panel surface finish including adhesives and sealants, are major design parameters universal to structural shear joints.

Open Hole Panels. Open hole panels with straight and countersunk holes are *not* shear connections but their analyses are particularly useful for treating multiple row joints. Idealized examples, with a centrally located, circular hole, uniformly loaded in uniaxial tension are shown in Figure 1.4. Three variations of the open hole are considered in this text: (i) a straight empty hole, (ii) a straight hole filled with a close fitting fastener shank [24] and (iii) an empty countersunk hole (see Chapter 7.3). Unlike shear connections, no load is applied to the panel at the site of the hole.

1.2 OUT-OF-PLANE DEFORMATION, MICROSLIP AND PLASTICITY

Out-of-Plane Deformation, Clamping, Interference and Friction. The loading of the open-hole panel and the attachment joint fastened with headless pins does not produce displacements or stress gradients in the out-of-plane directions provided the hole and pin diameters are uniform (straight hole). The gradients in a butt joint assembled with conventional fasteners are also usually negligible in the absence of lateral clamping. In these cases, 2-dimensional analyses are appropriate. The tilting of the lap joint fasteners, which is accompanied by the bending of the panels, and clamping produce significant stress gradients in the thickness direction; these increase the peak stress and shift its location to the panel interior surface. The tilting also produces contact pressure peaks under the fastener head edge. For

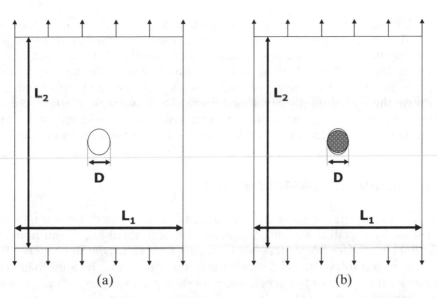

Figure 1.4. Schematic of open hole panels: (a) empty hole and (b) filled hole.

these reasons, the lap joint and the clamped butt joint must be treated as 3-dimensional problems. The analysis of shear joints is further complicated by friction and by interference at the fastener-panel interfaces. Ciavarella and Decuzzi [25] have devised a closed form solution for the 2-dimensional, frictionless, pin-panel contact problem for *elastically similar* material pairs without interference or clamping. However, analysts must resort to the finite element methods described in Chapter 13 to examine friction, clamping, arbitrary material pairs and the proximity of free edges even for 2-dimensional models of joints.

Microslip. The loading of a shear joint is accompanied by microscale-size, relative movements of the components at the contacting, fastener-panel and panel-panel interfaces. These movements, referred to here as microslip, contribute to the total deformation of the joint and affect the stress distribution. Microslip in structural joints is controlled by friction and is a highly non-linear type of deformation behavior. Consequently, even when the panels and fasteners of shear joints display linear elastic stress-strain behavior, the assembled joint will depart from linear behavior. The departures are evaluated in Chapter 5.6.

Plasticity. The onset of plastic deformation, either local yielding at stress concentrations or general yielding, also contributes to departures from linear elastic behavior. These departures are obtained at relatively low nominal stresses in view of the large stress concentrations. For example, a single rivet-row, aluminum butt joint (yield stress = 500 MPa) will begin to yield locally when the applied (gross section) stress exceeds 87 MPa. Above this nominal stress the concept of the SCF fails to describe stress elevation and elastic-plastic analyses are needed to define the local stresses and strains.

Cyclic Stress-Strain Behavior. Elastic-isotropic-plastic analyses can evaluate the strains during the first loading cycle. When the joint is subject to repeated cycles of loading, more sophisticated, elastic-kinematic-plastic analyses that account for the hysteresis loop are needed to define the steady state plastic strain amplitudes [26–28]. The strain amplitude is a crucial descriptor of fatigue, but its evaluation calls for the finite element modeling of the joint for multiple cycles of loading and unloading that precede the steady state. Such modeling presents difficult problems because it is computationally intensive and because the constitutive theory is not fully developed. Some elastic-isotropic-plastic and elastic-linear-kinematic-plastic analyses of joints are described in this book, but these were performed mainly to evaluate the effects of local plasticity on contact pressure and slip.

1.3 FASTENERS AND THE FASTENER LOAD

Generic types of fasteners are illustrated in Figure 1.5. They include rivets, whose heads are formed in place by upsetting a blank, and bolts with preformed heads and retaining nuts. Rivets and bolts may have protruding heads, or countersunk heads. The fastener head and shank play important roles. The head allows the fastener to clamp the joint together and its rigidity limits local bending of the panels and important peak stresses. At the same time, the benefits of clamping in the panel hole are partly offset by the stress concentration in the panel underneath the edge of the rivet head. This trade-off can be an important factor in design. The shank must support tensile stresses which can be very high when load transfer relies on high levels of lateral clamping. The compressibility of the shank is reduced adjacent to the more massive head; this introduces gradients of contact pressure and stress in the thickness direction for fasteners installed with interference.

The fastener load, Q, is the load transmitted to the pin, rivet or bolt holding the joint together. For single fastener-row joints, Q is the product of the gross section stress, σ, the panel thickness, t, and the fastener spacing or pitch, P_1 (see Figure 1.2):

$$Q = \sigma t P_1 \qquad\qquad (1.1)$$

In multiple row joints, the load Q is distributed among all the rows, but *not evenly!* The fasteners in the leading and lagging rows of holes always support a larger portion of Q, with intermediate rows carrying progressively smaller portions. This is described more fully in Chapters 2.6 and 17.

1.4 THE BEARING, CLAMPING AND ADHESIVE MODES OF LOAD TRANSFER

The fastener load, Q, can be transmitted from one panel to the other by one or a combination of three modes of load transfer.

Bearing Mode. In the bearing mode, Figure 1.6a, load transfer proceeds mainly at the panel hole-fastener shank interface. Panel 1 presses against one end of the fastener shank (at A), and the other end of the shank presses against the opposite panel (at B) stabilizing panel 2. The bearing mode produces high levels of contact pressure and tensile stress at the panel hole periphery. A detailed analysis of load transfer by the bearing mode in a lap joint is presented in Chapter 6.

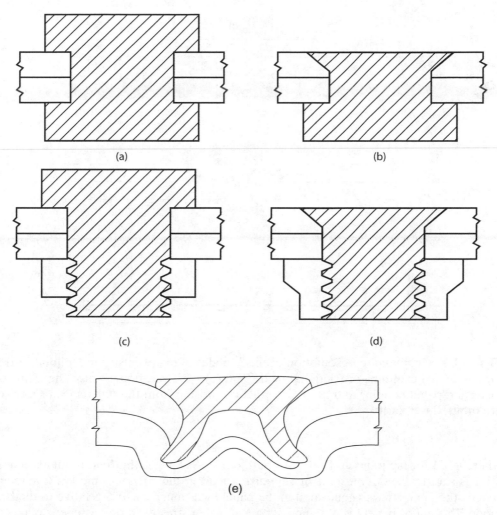

Figure 1.5. Examples of fasteners used in shear connections: (a) rivet with standard head, (b) countersunk rivet, (c) standard bolt, and (d) countersunk aircraft-type bolt, (e) self-piercing rivet.

Clamping Mode. The clamping mode, also referred to as the *frictional, slip resistant* or *slip critical* mode, is illustrated in Figure 1.6b. A clamping force, Q_C, is produced by torquing the bolt or upsetting the rivet. This force is also called the bolt tension. The resulting contact pressure at the panel-panel interface under the fastener head (at A, Figure 1.6b) produces the friction traction, μQ_C (μ is the coefficient of friction) which transfers load from panel 1 to panel 2. In the clamping mode the fastener hole periphery remains largely unstressed. The maximum fastener load that is supported by friction is:

$$Q = \varphi\,\mu\,Q_C \qquad\qquad (1.2)$$

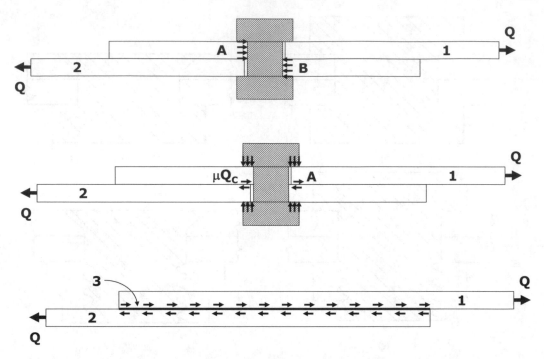

Figure 1.6. Schematic representation of the 3 modes of load transfer in lap joints: (a) bearing mode, (b) clamping or frictional mode, and (c) adhesive mode. Note: The tilting of the fasterner, panel bending and attending tractions arising from the unbalanced shear couple are omitted for clarity.

where $\varphi = 1$ for lap joints and $\varphi = 2$ for butt joints[1]. As long as this load is not exceeded, the joint is said to be *slip-resistant* or supporting load in the *frictional* mode[2]. Clamping can reduce the peak stresses generated in the panel by as much as 70% relative to the bearing mode. This major benefit is obtained at cost of higher stresses in the fastener shank.

Combined Clamping-Plus-Bearing Mode. Beyond the load carrying capacity of the clamping mode (Equation 1.2), load transfer is accomplished by the combination of an essentially fixed clamping component and a growing bearing component. However, the benefits of clamping—reduced peak stresses—persist even after the bearing component becomes dominant.

Adhesive-Plus-Bearing Mode. The adhesive mode employs shear tractions supported by the adhesive layer, 3 in Figure 1.6c, to transfer load from panel 1 to panel 2. When combined

---

[1] The 2-fold difference in the value of φ arises because clamped butt joints develop shear tractions on two surfaces compared to one for lap joints.

[2] For reasons discussed in Chapter 3.1, the panel does not slip into the bearing mode as soon as this value of Q is exceeded.

with rivets and bolts, the adhesive reduces the fastener load and the peak stresses generated at the fastener holes. The reduction in fastener load is directly related to the stiffness of the adhesive. However, even low modulus sealants, whose main function is to restrict moisture penetration, can reduce peak stresses substantially. This is examined in Chapter 2.5.

1.5 STRESS DESCRIPTORS FOR FATIGUE

Shear joints subjected to repeated cyclic loading are vulnerable to contact fatigue failure, which is initiated in the bulk. The distributions of the stresses, contact pressure and microslip, in the vicinity of the fastener hole and at the interfaces, are complex and non-uniform. The peak values of stress amplitude and mean stress are critical for forecasting the fatigue life of the joint. Cyclically loaded joints also experience fretting, the to-and-fro rubbing (cyclic microslip) of contacting panel-panel and panel-fastener interfaces. Fretting can produce changes in the surface topography; it can promote wear and initiate fatigue failure, affecting the cyclic life of joints in special cases.

The Stress Concentration Factor (SCF). The peak tensile stress, σ^*, is conveniently described by the stress concentration factor (SCF) defined as the ratio of the peak tensile stress-to-nominal (gross section) stress: $SCF = \sigma^*/\sigma$. The SCF of the empty, open-hole panel, $SCF = 3$, is widely known [29]. Figure 1.4 illustrates that the hole is distant from the site of load application. Elevated stresses arise from the small load that is redistributed locally when the panel is perforated. The loading of shear joint panels is distinctly different. Here the fastener hole *is* the site of load application. The combination of large fastener load applied to a small area produces even larger local stresses; the SCFs of single fastener-row shear joint panels can be 2x to 4x the SCF-value of the open hole. The magnitudes and locations of the SCFs of shear joint panels depend on the mode of load transfer (see Table 1.1) and many other factors. This topic is dealt with more fully in Chapters 2 and 3. The varia-

TABLE 1.1. STRESS CONCENTRATION FACTORS, SCFS, FOR TENSILE LOADED, ALUMINUM, OPEN-HOLE PANELS, AND SINGLE RIVET-ROW LAP JOINTS

MODE OF LOAD TRANSFER	SCF	ANGULAR LOCATION[1]	Finite Element Model or Ref.
OPEN HOLE PANEL L1 , L2 >>D			
empty hole	**3.00**	0°, 180°	(29)
filled hole	**2.1**	0°, 180°	1H-5
filled hole	**2.85**	22°, 158°	1H-5
LAP JOINT, $P_1/D=5$, $D/t=4$			
Bearing	**6.1**	-3°, 183°	4S-1
Clamping	**4.5**	90°	
Bearing + Adhesive	**5.1**[2]	1°, 181°	4SA-5

1. The convention for angular location is given in Figure 2a.
2. SCF calculated for a low modulus sealant.

tions of contact pressure, microslip and tangential stress at joint interfaces are described in Chapter 3.4.

Fretting Factors. To quantify the possible influence of fretting wear and fretting fatigue, two fretting factors (F_1 and F_2) have been reported in the literature [30,31]. The fretting wear factor, F_1, is related to the depth of the wear scar. The fretting fatigue factor, F_2, is an approximate expression of the integrated effects of contact pressure, microslip and the bulk tangential stresses. Both parameters were empirically derived to model the influence of fretting on the reduction in fatigue life in turbine engine dovetail joints; they have been used on occasion to design these components. Values of F_1 and F_2 and the severity of fretting fatigue and wear of shear joints are assessed in Chapter 4.5. Discussion of the influence of fretting and the reliability of these parameters is covered more fully in Appendix C.

SHEAR JOINTS IN THE BEARING MODE

2.1 STRESS DESCRIPTORS FOR BEARING MODE

In the bearing mode, a relatively large fastener load, Q, is applied to the small area of contact between the fastener shank and the panel. The area corresponds approximately with the projected area of the shank, $A = D_S t$ (where D_S is the diameter of the shank and t is the panel thickness). In single fastener row joints, the concentration of load produces peak local stresses, σ^*, of order Q/A:

$$\sigma^* = f(Q/A) = f(Q/D_S t) = f(\sigma P_1/D_S), \text{ and} \qquad (2.1)$$
$$SCF = \sigma^*/\sigma = f(P_1/D_S) \qquad (2.2)$$

In addition, the stresses are affected by μ, the coefficient of friction, which governs microslip at the joint interfaces and, in the case of lap joints, by the fastener shank diameter-to-panel thickness ratio, D_S/t, which affects panel bending. The geometry of the fastener head and its rigidity also play a role because the clamping forces applied by the fastener head reduce panel bending under the head. The number of fastener rows also has a large effect on joint response because, Q_i, the actual load transmitted to the fasteners of a particular row, is a fraction f_i, of the (total) fastener load: $Q_i = f_i Q$. Finally, the residual stresses attending shank-hole interference have major effects. These are described in Chapter 3.3. It follows that the mechanical response of the joint—including the joint compliance, rivet tilt, local stresses, strains, and displacements, the local contact pressure and slip, and the state of the fasteners—depends on the following variables:

$$\text{Joint response} = f(f_i \sigma P_1/D_S, D_S/t, \mu, \text{ interference, fastener head geometry}) \qquad (2.3)$$
$$SCF = f(f_i P_1/D_S, D_S/t, \mu, \text{ interference fastener head geometry}) \qquad (2.4)$$

The material properties of the panels, fasteners, and sealant or adhesive at the joint interfaces, the type of loading—uniaxial or biaxial—and the proximity of free edges are additional variables.

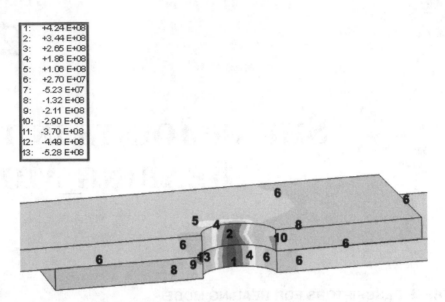

1:	+4.24 E+08
2:	+3.44 E+08
3:	+2.65 E+08
4:	+1.86 E+08
5:	+1.06 E+08
6:	+2.70 E+07
7:	-5.23 E+07
8:	-1.32 E+08
9:	-2.11 E+08
10:	-2.90 E+08
11:	-3.70 E+08
12:	-4.49 E+08
13:	-5.28 E+08

Figure 2.1. The σ_{11}-stress contours in the thickness direction of a 3D model of a butt joint showing the absence of a significant stress gradient. The cover panel and half the center panel of a single rivet-row butt joint fastened with standard head rivets are shown. The applied stress is 60 MPa. (Model 3S-1).

2.2 BUTT JOINT

This section presents results of 2- and 3-dimensional analyses of a 1:2:1[1], aluminum butt joints in the bearing mode. The 2-dimensional analyses preclude out-of-plane gradients and can only represent a joint secured with a headless pin. The 3-dimensional models treat a joint fastened with rivets or bolts with standard heads and allow out-of-plane bending of the rivet shank and panels and rivet tilt. In spite of these differences, the results of the two sets of analyses are similar. The σ_{11} stress contours in Figure 2.1 illustrate that out-of-plane gradients are virtually absent. The SCF-values listed in Table 2.1 for the 2-dimensional model and the center panel of the 3-dimensional model, SCF = 6.4 and 6.3, respectively, are nearly the same. The SCF-value at the faying surface of the cover plates, SCF = 6.4, is only 1.6% higher; fastener bowing is ~0.2° when σ = 60 MPa The findings confirm that the out-of-plane bending of butt joints is usually negligible and that results obtained with 2–dimensional analyses, such as those in Chapter 7.2, are useful approximations for butt joints in the bearing mode[2].

Figure 2.2a shows the angular conventions used throughout this book. Features of the in-plane σ_{11} stress distribution in the vicinity of a filled open hole are illustrated in Figure

---

[1] 1:2:1 are the relative thickness values of the butt joint cover, center and cover panels.

[2] Out-of-plane variations in stress are obtained in clamped joints (see Chapter 3.2).

TABLE 2.1. STRESS CONCENTRATION FACTORS, SCFS, FOR ALUMINUM ATTACHMENT JOINTS AND BUTT JOINTS IN THE BEARING MODE

ANALYSIS	GEOMETRY AND FRICTION		SCF	ANGULAR LOCATION [1]	Finite Element Model or Ref.
Narrow joint with single, headless fastener					
	$L_1/D_S=5$	$\mu=0$	6.1	2°, 178°	3P-13
	$L_1/D_S=5$	$\mu=0.2$	6.3	4°, 176°	3P-16
2-D analysis	$L_1/D_S=5$	$\mu=1.0$	8.2	4°, 176°	3P-18
	$L_1, L_2=\infty$	$\mu=0.2$	4.5	4°, 184°	3P-6
Wide, multi-fastener joint with single row of headless fasteners					
2-D analysis	$P_1/D_1=5$	$\mu=0.2$	6.4	0°, 180°	3P-1
2-D analysis	$P_1/D_C=8$	$\mu=0.2$	7.9 [2]	0°, 180°	3P-1
3-D analysis, cover panel, faying surface	$P_1/D_S=5$	$\mu=0.2$	6.4	0°, 180°	3S-1
3-D analysis, center panel, faying surface	$P_1/D_S=5$	$\mu=0.2$	6.3	0°, 180°	3S-1

1. The convention for angular location is given in Figure 2a.
2. SCF value estimated using the Seliger SCF dependence on P/D.

2.2b. The in-plane σ_{11}-stress contours for the butt joint are shown in Figure 2.2c. It can be seen that the bearing mode produces intense contact pressures and compression on both sides of the panel-fastener shank interface in a 120°-sector centered on $\theta = 270°$. The panel and fastener are unloaded on the opposite side of the hole in a 120°-sector centered on $\theta = 90°$, where a gap between the fastener and the panel develops (Figure 2.2a). A stress concentration with peak tensile stresses forms in the panel adjacent to the hole periphery near the $\theta = 0°$- and $\theta = 180°$-locations (Figure 2.2c). The peak tensile stress displays a nearly linear dependence on the nominal stress and a small amount of hysteresis related to slip and friction as illustrated in Figure 2.3a; the corresponding SCF is essentially constant (Figure 2.3b). The peak stress and SCF at the 0° and 180°-locations are here referred to as "net section" peak stress and "net section" SCF. The peak stress at the $\theta = 90°$, $r = 1.2$ r_H[3] and the corresponding SCF are referred to as the "gross section" peak stress and "gross section" SCF (Figure 2.3a and 2.3b). The net section SCF is larger than the gross

---

[3] Where r is radial distance measured from the center of the fastener hole and r_H is the local radius of the stressed hole.

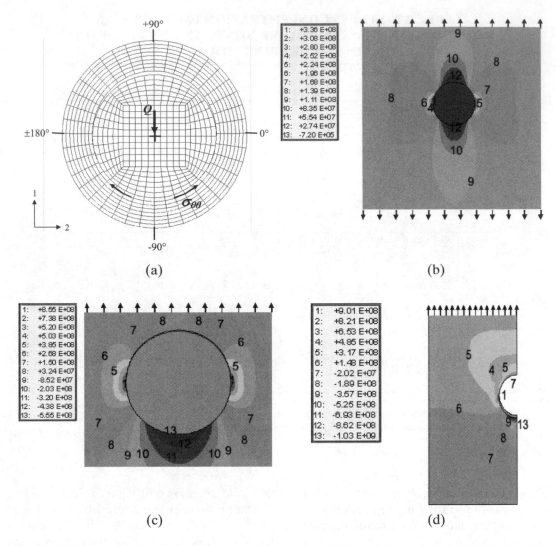

(a)   (b)

(c)   (d)

Figure 2.2. Displaced mesh and σ_{11}-stress contours on in-plane sections of the panel adjacent to the hole edge: (a) example of the displaced mesh for wide, single fastener row butt joint showing convention for angular location, $P_1/D_S = 5$, (b) filled open-hole panel, SCF = 3 (1H-5), (c) wide, single fastener-row butt joint, $P_1/D_S = 5$, SCF = 6.4, (3P-1), and (d) wide, single rivet-row lap joint, $P_1/D_S = 5$, $D_S/t = 4$, SCF = 6.1 (4S-1), $\sigma_{nom} = 125$ MPa.

section SCF for the bearing mode of load transfer, but the opposite is true for lap joints in the clamping mode. This is important because the location of the dominant SCF determines the mode of joint failure.

Results for contact pressure, microslip and the tangential stresses at the shank-panel interface of butt joints are presented in Chapter 3.4.

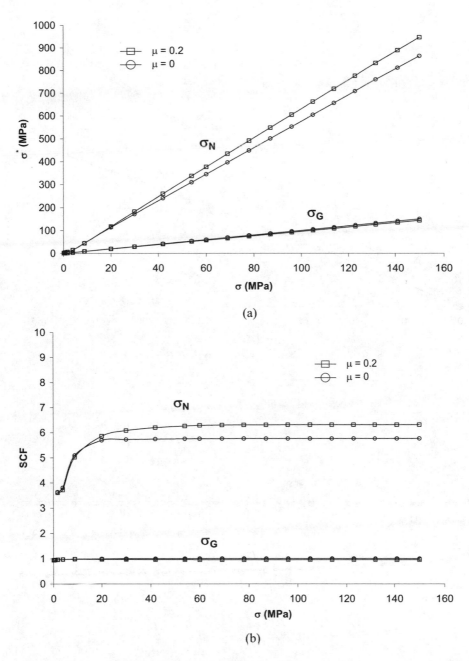

Figure 2.3. Variations with applied stress of the .peak, net section stress, gross section stress and corresponding SCFs for a wide, single fastener row, aluminum butt joint and lap joint panel in the bearing mode, $P_1/D_S = 5$, $D_S/t = 4$: (a) and (b) butt joint. (continues)

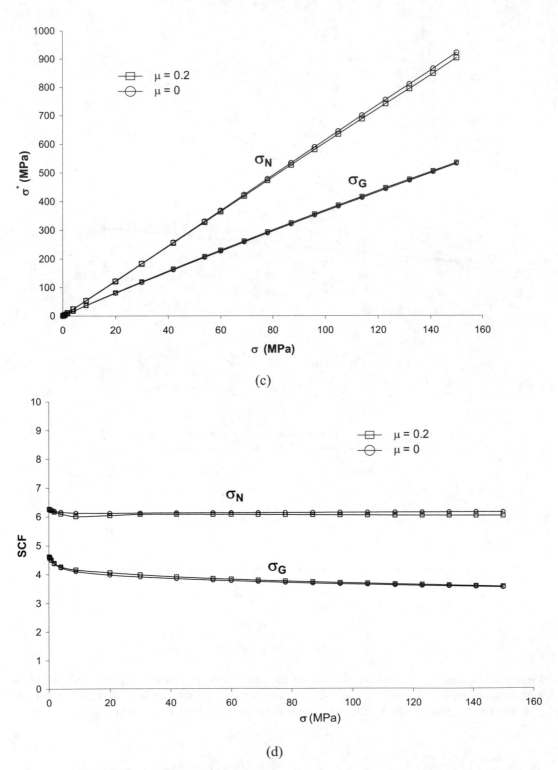

(c)

(d)

Figure 2.3. (continued) (c) and (d) lap joint. Models (3S-1) and (4S-29), (22).

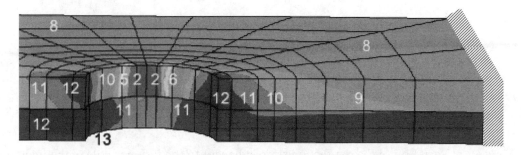

Figure 2.4. Stress contours generated in a wide, 2-row aluminum doubler ($P_2 = P_1$, $P_1/D_s = 5$, $D_s/t = 4$, $\mu = 0.2$ and $\sigma = 150$ MPa) fastened with standard head rivets and no clamping (Model 7S-1). The contours illustrate that stress variations in the thickness direction are negligible.

2.3 DOUBLER

The stress contours for a 2-row doubler ($P_2 = P_1$, $P_1/D_S = 5$, $D_S/t = 4$) are reproduced in Figure 2.4. They illustrate that stress distribution, like that of the butt joint, is essentially 2-dimensional with negligible variation in the thickness direction. According to the analysis of this particular 2-row doubler (Chapters 17.2, 17.3 and Table 17.4), about 10% of the load applied to the main panel is diverted to the doubler panel[4] while a stress concentration, SCF $\cong 2.8$, is generated in the main panel. This SCF is in agreement with the value, SCF $= 2.9$, derived from the finite element model and Figure 2.5. This means that the 10% reduction in the stress carried by the main panel is offset by the 290% increase in stress at the main panel fastener hole edge. With a 4-fold increase in the fastener row separation, $P_2 = 4P_1$, a larger stress reduction, about 20%, and SCF $\cong 3.2$ are obtained. Benefits are realized when the doubler straddles a detail in the main plate with a large stress concentration. For example, the 20% reduction in stress afforded by the doubler is equivalent to reducing a detail SCF $= 6$ to SCF $= 3.8$, this at the cost of the SCF $= 3.2$ concentration introduced at the main panel fastener holes. Larger benefits are possible with 4- and 6-row doublers (see Table 17.4).

---

[4] For the following values appropriate for this doubler: $k/K = 0.18$ and $K_S/K \cong 1.5$, the fastener load fraction, $Q_1/Q \sim 0.1$.

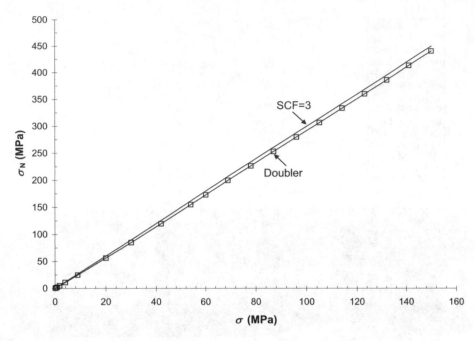

Figure 2.5. Variation of the net section stress adjacent to the fastener hole in the main panel of the wide, 2-row doubler referred to in Figure 2.4 (Model 7S-1).

2.4 LAP JOINT

Single Rivet-Row. Results obtained with 3-dimensional analyses of lap joints differ from those for butt joints in one important respect: the effects of out-of-plane bending of lap joints are not negligible. Rivet tilt and out-of-plane bending of two lap joints are illustrated in Figures 2.6a and 2.6b. Both the tilt and bending increase with fastener load. The σ_{11}-stress contours are superimposed on the deformed meshes in Figures 2.6c and 2.4d. The angular variation of the in-plane, tangential, $\sigma_{\theta\theta}$–stress (at the edge of the hole) at the panel top surface ($z = 0$), on the mid-plane ($z = 0.5t$) and the bottom surface ($z = t$) are shown in Figure 2.7. The variations with the thickness direction are produced by the out-of-plane panel bending displayed by the distorted meshes in Figure 2.6. The panel bending displaces the highest stresses and the SCF to the interior face of both panels. An important consequence is that fatigue cracks are likely to initiate at the SCF-location where they are hidden from direct observation by both the rivet head and the panel[5]. This complicates the task of non-destructive evaluation (NDE) of fatigue damage.

Figures 2.3c and 2.3d describe the variation with nominal stress of the peak σ_{11}-stresses and SCFs at the depth $z = t$, the panel faying surface for the standard rivet head. The nearly linear variation with nominal stress at the net section and gross section locations and the

---

[5] When cracks initiate at the faying surface of the back panel, the panel-panel interface acts as an additional obstacle to NDF detection.

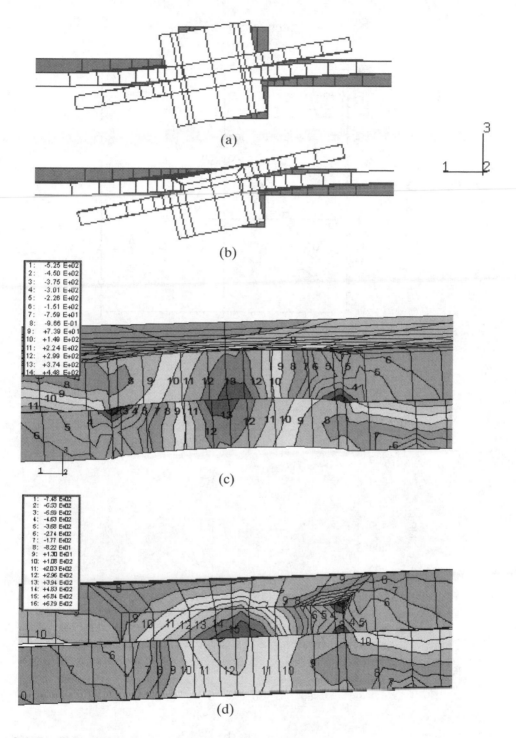

Figure 2.6. Profiles of the displaced meshes and corresponding σ_{11}-stress contours in the vicinity of the panel holes (with the fasteners not shown) for 2, wide, aluminum, single rivet-row lap joints, $P_1/D = 5$, $D/t = 4$, nominal stress, $\sigma_{nom} = 125$ MPa: (a) standard rivet head with (b) corresponding stress contours (4S-1), and (c) $100°$-countersunk head with (d) corresponding stress contours (4C-1). The shaded portions of the distorted meshes indicate the initial configuration. The displacements of the meshes are magnified 3x.

Structural Shear Joints

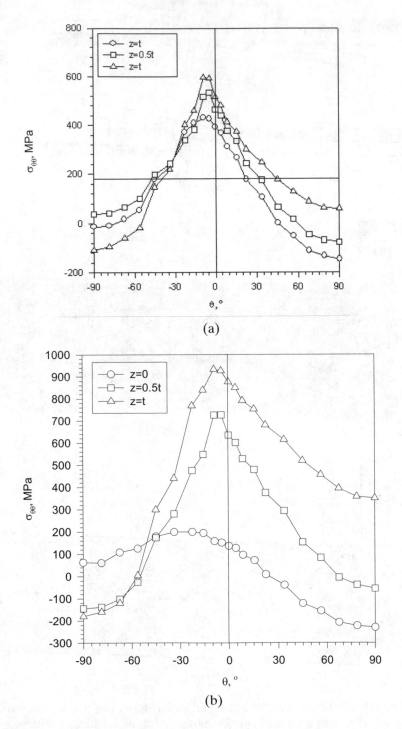

Figure 2.7. The angular and with-depth variation of the tangential stress, adjacent to the fastener hole of wide, aluminum, single rivet-row lap joint panels ($P_1/D = 5$, $D/t = 4$, = 125 MPa): (a) standard rivet heads (4S-1), and (b) countersunk heads (4C-1).

TABLE 2.2. PROPERTIES OF WIDE, TENSILE LOADED, ALUMINUM, 1 AND 2 RIVET-ROW LAP JOINTS IN THE BEARING MODE

LAP JOINT	GEOMETRY	EXCESS COMPLIANCE , m/GN	RIVET TILT degrees	SCF	LOCATION[a] θ r z	F.E. MODEL
1. 1 Rivet-Row	1.1 Std. aluminum rivet head	27.5	3.6	**6.1**	-3°, 183° r_H t	4S-1
P_1/D=5, D/t=4	1.2 Aluminum rivet 100°-csnk to 1/2 panel thickness	34.4	4.2	**9.8**	-3°, 183° r_H t	4C-1
	1.3 Same as 1.2 but with steel rivet	28.3	4.0	**8.8**	-1°, 181° r_H t	4C-2
	1.4 Aluminum rivet 100°-csnk to full panel thickness	41.2	4.1	**12.1**	-6°, 186° r_H t	4C-6
2. 1 Rivet-Row	2.1 Std. aluminum rivet head			**7.5**[d]		4S-1[d]
P_1/D=8, D/t=4	2.2 Aluminum rivet 100°-csnk to 1/2 panel thickness			**12.1**[d]		4C-1[d]
	2.3 Aluminum rivet100°-csnk to full panel thickness			**14.7**[d]		4C-6[d]
3. 2 Rivet-row	3.1 Std. aluminum rivet head	7.5	1.7[b], 1.7[c]	**4.4**[b], **3.1**[c]	-2°[b], -3°[c] r_H t	5S-1
P_1/D=5, D/t=4	3.2Aluminum rivet 100°-csnk to 1/2 panel thickness	10.3	1.5[b], 1.9[c]	**6.1**[b], **4.5**[c]	-3°[b], -5°[c] r_H t	5C-1

a. Location of SCF given in polar coordinates. Convention for angular location, θ, is given in Figure 2a. The quantity, r_H, is the local radius of the stressed fastener hole; z is the depth measured from the outer surface of the panel.
b. Leading row rivets
c. Lagging row rivets
d. SCF values estimated using Seliger (32) SCF dependence on P_1/D

dominant SCF are similar to those found for the butt joint. The gross section SCF of the lap joint exceeds that of the butt joint. The SCFs of a number of lap joints are listed in Table 2.2. Other features of the loaded lap joint including the excess compliance, rivet tilt, and stresses and displacements at key locations are tabulated in Chapter 12.

Both the amount of bending and its distribution and the lateral constraint imposed by the rivet head affect the stress concentration. This is illustrated by the results for a wide aluminum lap joint (P_1/D = 5, D/t = 4, σ = 125 MPa) with standard rivet heads which displays a rivet tilt of α = 3.7°. The tilt, which is ~10x larger than the bowing of a butt joint fastener, produces significant panel bending which elevates the tensile stresses at the panel faying surface. Yet, the net section stress concentration of the lap joint, SCF = 6.1, is less than the value obtained for a comparable butt joint, SCF = 6.4, which hardly bends at all. The muted affect of bending is related to the distribution of bending. The lateral constraint produced by the relatively massive and rigid rivet head limits the amount of panel bending directly under the head where the net section SCF is located. The bulk of the bending is shifted to the unclamped region outside the head. In addition, the contribution of bending in a lap joint is offset by the frictional component of load transfer that reduces the force on the rivet by 10% (see Chapter 6). In contrast, the countersunk rivets display a tilt, α = 4.2° (for α = 125 MPa), and the countersunk panel a net section SCF = 9.8. The 14% higher tilt angle and attending increase in panel bending account for a small part of the 61% increase in the SCF. Two other factors are at play. First, the main part of the bearing load is transmitted to the countersunk rivet by the non-countersunk faces at the lower half of the panel. As a result, the tangential tensile stresses are skewed toward the lower half of the panel at depths between 0.5t < z < t, and this adds to the concentration of stress near the interior panel face produced by panel bending (compare Figures 2.7a and 2.7b). Secondly, the less massive and more compliant countersunk head allows more bending at the critical location under the head, as shown by the greater variation of the peak circumferential stress with thickness in Figure 2.7b. These findings illustrate that both the shape of the fastener and the rigidity of the fastener head affect the panel SCF.

Double Rivet-Row. The stress contours and distortions of a two rivet-row lap joint fastened with standard head rivets are illustrated in Figure 2.8. It follows from the symmetry that the fastener loads applied to the two rows are identical and half the value of the single row joint described in the previous paragraphs. The net section SCFs of the single and double row joints in Table 2.2 illustrate two generalizations that apply to multiple-row joints:

i. The local stresses at the leading row do not diminish in proportion to either the number of rows or the fastener load. In this case, doubling the number of rows reduces the SCFs at the leading row by only 28% and 38% for the standard and countersunk rivets, respectively.
ii. The local stresses at the leading row are higher than at the lagging row. In this case the stresses at the leading row are 42% and 36% higher at the leading row for the standard and countersunk rivets, respectively.

The origins of these unexpected variations of fastener load and the SCF in multiple-row joints are dealt with in Chapter 2.6.

Influence of the Pitch/Diameter- and Diameter/Thickness-Ratio. Local stresses in the fastener and panel depend only on Q, the fastener load, as long as the separation of fasteners is large and the overlap of stresses from adjacent fasteners is negligible. In that case, the SCF varies linearly and directly with the P_1/D_s –ratio (see Equation (2.2)):

$$SCF = \psi \, (P_1/D_s) \tag{2.5}$$

where the value of ψ is influenced by other geometric variables. For very small fastener separations, $P_1/D_s < 2$, the SCF will be dominated by the net section (the reduced load bearing cross section of the panel) and will vary inversely with P_1/D_s. For intermediate values of P_1/D_s, a transition from one to the other dependence is to be expected. Crude estimates of the 3 dependencies are offered in Figure 2.9 along with experimental results and the value derived from calculations.

Seliger [32] measured the fatigue lives of single rivet-row lap joints with different rivet spacings in the range $4 < P_1/D_S < 12$, and compared these with the fatigue lives of unperforated samples of the panel material. He then deduced values of the (net section) SCF of the joints from the ratios of the stress levels that produced comparable lives. The Seliger results plotted in Figure 2.9 correspond with a fatigue life of $N = 10^7$ cycles, conditions least likely to involve local yielding. His values and the calculated value for $P_1/D_S = 5$ are in accord. However, they do not conform with Equation (2.5), possibly for two reasons. First, the value of the SCF measured for $P_1/D_S = 12$ may be understated because of the onset of yielding (See Appendix B). Secondly, the range $2 < P_1/D_S < 6$ probably corresponds with the transition. More work is needed to evaluate ψ and define the variation of the SCF in the transition regime.

The effects of panel thickness on single rivet-row lap joints have been evaluated with finite element calculations by Fongsamootr (33); the variations of SCF with D_S/t are shown

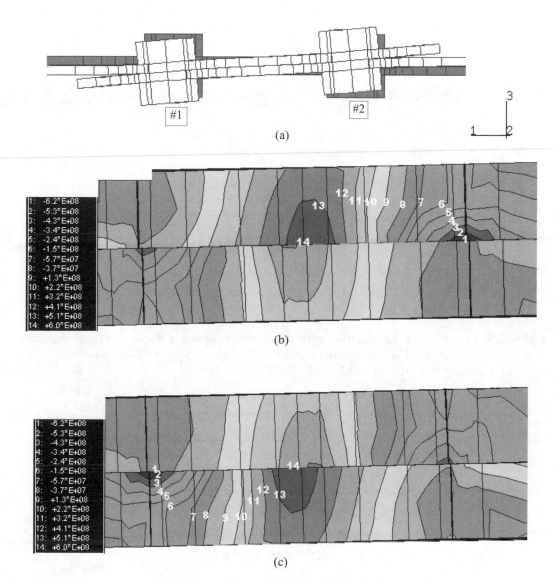

Figure 2.8. Profiles of the displaced mesh and the σ_{11}-stress contours for an aluminum, double rivet-row lap joint, $P_1/D = 5$, $D/t = 4$, nominal stress, $\sigma_{nom} = 125$ MPa (5S-1): (a) displaced mesh; displacements magnified 3x; shaded areas show initial configuration, (b) stress contours for hole #1, (c) stress contours hole for #2.

in Figure 2.10. The two sets of results (Figures 2.9 and 2.10) make it possible to generalize findings obtained for a particular, P_1/D_S- and D_S/t-lap joint configuration. Although the Seliger's P_1/D_S-results were obtained for lap joints, they offer approximations for doublers, butt and attachment joints as well. The D_S/t dependence does not apply to butt and attachment joints display little or no panel bending.

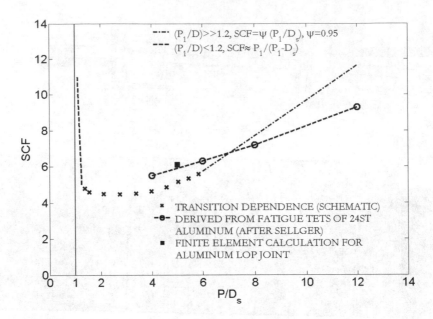

Figure 2.9. Estimates of the variation with the pitch-to-diameter Ratio, P_1/D_S, of the stress concentration factor (SCF) for aluminum, single rivet-row lap joint panels including experimental results of Seliger [32] and Model (4S-1).

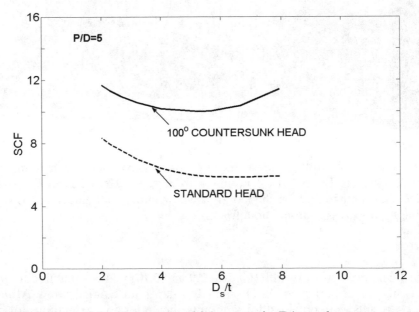

Figure 2.10. Influence of the diameter-to-thickness ratio, D/t, on the stress concentration factors for aluminum, single row lap joint panels, $P_1/D = 5$ with standard head rivets after Fongsamootr [33].

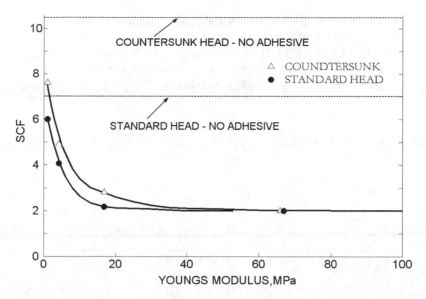

Figure 2.11. Influence of adhesive stiffness on the stress concentration factor (SCF) for aluminum, narrow, single rivet lap joint panels, $L_1/D_S = 5$, $D_S/t = 4$ (6-8,21). The calculations are for standard head rivets with a 300 ?m-adhesive layer (Models 3S-8, 3SA-6, 19, 20, and 21, and for 100°-countersunk head rivets with a 180 μm-adhesive layer (Models 3C-9, 3CA-6).

2.5 LAP JOINTS IN THE ADHESIVE-PLUS-BEARING MODE

This section describes findings for lap joints fastened with both rivets and adhesive[6] where load transfer proceeds by both the bearing and the adhesive mode [21]. Results of finite element analyses of lap joints fastened exclusively with adhesive are summarized in Chapter 16 [6,21].

The installation of an adhesive at the interfaces of a pinned or bolted joint offers important mechanical benefits. Relatively stiff adhesives dramatically increase load transfer directly across the panels, thereby reducing the bearing load on the fastener, the amount of tilting and bending, and the peak stresses and SCF generated in the panel. Figure 2.11 describes the relation between the panel net section SCF and the stiffness of hypothetical adhesives. These calculations assume that the adhesive remains intact: does not fail or separate from the panel interfaces. This requires that the adhesive supports peak tensile and shear stresses that increase with the stiffness of the adhesive. Examples of the magnitude of these stresses are listed in Table 2.3.

Polymer sealants, which are installed primarily to shield the joint interfaces from corrosion, are very compliant and may possess an elastic modulus as low as $E = 1$ MPa, and an equally low, pressure dependent shear modulus. This stiffness corresponds with the data points at the extreme left of Figure 2.11. Nevertheless, low modulus sealants can improve

---

[6] The models treat joints with adhesive at both the panel-panel and fastener-panel interfaces.

**TABLE 2.3. PEAK TENSILE AND SHEAR STRESSES
SUPPORTED BY A 180 μ-THICK LAYER OF ADHESIVE
INSTALLED AT THE INTERFACES OF A 30.6 mm-WIDE,
SINGLE RIVET-ROW, ALUMINUM LAP JOINT,
$L_1/D = 5$, $D/t = 4$**

After Fongsammootr (7)

ADHESIVE TENSION MODULUS[a], MPa	PEAK TENSILE STRESS, MPa	PEAK SHEAR STRESS, MPa	F.E.MODEL
1.1	0.65	0.51	4SA-6
4.4	1.18	1.41	4SA-19
17.6	1.82	2.95	4SA-20
70.4	3.22	5.08	4SA-21

a. Combined with a pressure sensitive shear modulus

**TABLE 2.4. PERCENT CHANGES IN THE MECHANICAL
BEHAVIOR PRODUCED BY A 180 μ-THICK LAYER OF
SEALANT[a] INSTALLED AT THE INTERFACES OF RIVETED,
SINGLE-ROW, ALUMINUM LAP JOINTS, $P_1/D = 5$, $D/t = 4$**

After Fongsammotr (7) and Kamnerdtong (8)

F.E. MODEL	4S-8, 4SA-5	4S-10, 4SA-7	4C-9, 4CA-25
RIVET HEAD	Standard	Standard	100°-countersunk
PANEL THICKNESS, mm	1.53	1.00	1.53
NOMINAL STRESS, MPa	65	65	60
Excess compliance	+14	.+12	.+14
Rivet tilt	-2°	-7°	-2°
In-plane slip	+42	+52	+4
Out-of-plane slip	+700	+514	+126
Panel-shank contact pressure	+16	+21	+28
Lateral separation of panels	-40	-57	-57
SCF	-16	-18	-28

a. Desoto PR-1776 Type B-2 sealant, Youngs Modulus, E=1 MPa

mechanical performance significantly. The computed changes in the mechanical response of single rivet-row lap joints are summarized in Table 2.4. These show that the sealant, acting like a lubricant, increases joint compliance and the in-plane and out-of-plane movements of the panels. The SCFs are reduced by 16% to 35% along with lateral separation of the panel ends, fastener tilt and panel bending. These changes are accompanied by 10-fold increases in the joint fatigue life, described in Chapter 4.3.

TABLE 2.5. VALUES OF THE FASTENER LOAD RATIOS, f,
AND THE SCFS OF DIFFERENT, MULTIPLE ROW JOINTS
WITH ALUMINUM PANELS AND RIVETS, P/D = 5, D/t = 4
AND P_2/P_1 = 1, DERIVED FROM THE 2 SPRINGS MODEL
AND THE SUPERPOSITION METHOD

JOINT TYPE	k / K	ROW 1*		ROW 2		ROW 3		ROW 4	
		f	SCF	f	SCF	f	SCF	f	SCF
1-row lap joint, std. rivet head (obtained from analysis 4C-1)		1	6.1						
2-row lap joint, std. rivet head	0.2	0.5	4.3	0.5	3.1				
3-row lap joint, std. rivet head	0.2	0.36	3.8	0.28	2.6	0.36	2.2		
4-row lap joint, std. rivet head	0.2	0.28	3.5	0.22	2.6	0.22	2.0	0.28	1.7
4-row lap joint, std. rivet head, and 1% interference	0.4	0.31	3.6	0.19	2.4	0.19	1.9	0.31	1.9
4-row lap joint, csk. rivet head	0.2	0.28	4.5	0.22	3.4	0.22	2.9	0.28	2.7
4-row doubler, std. rivet head	0.2	0.19		0.06		0.06		0.19	
Panel			3.2		2.3		2.3		3.2
Doubler			1.2		0.8		0.8		1.2
8-row butt joint, pinned	0.5	0.33	3.8	0.17	2.3	0.17	1.9	0.33	2.1

* Leading row of holes

2.6 MULTIPLE ROWS OF FASTENERS

Additional load carrying capacity or a reduction in local stresses is achieved by adding one or more rows of fasteners to shear joints. However, progressively smaller benefits are realized with increasing number of rows. This is related to the distribution of fastener load and the contribution of the bypass load to the SCF [24,34].

Estimating Fastener Load. The load applied to one panel of a multiple row joint is progressively transferred to the facing panel(s) by successive rows of fasteners. The heavily loaded panel incrementally sheds load, while the initially unloaded, facing panel, is incrementally loaded. As a result of this sequence, there are disparities in the amounts of load supported by facing strips of panel between the same fastener rows. The load disparities are accompanied by mismatch in the amounts of extension and Poisson's contraction displayed by the facing strips [34]. To maintain continuity of the fasteners, the sum of the deformation produced by: (1) the fastener-fastener hole complex, and (2) the intervening panel strip, must be identical for facing strips between the same two rows. The load and deformation disparities together with the compatibility requirement affect the fastener loads transmitted by the individual rows as well as the variations of fastener load within a row near the free edge (see Chapter 2.7).

One-dimensional spring models, that account for the compatibility requirement and the interactions between the fastener-fastener hole combination and the intervening panel strip, have been used by Swift [24] and Muller [34] to estimate fastener loads. A 2-springs model, devised by the authors is described in Chapter 17. The stiffness of one spring, k, characterizes the fastener-fastener hole response; the stiffness of the other, K, the response of the intervening panel strip. The models define f_i, the fractions of the total fastener load acting at each fastener row, as a function of the stiffness ratio, k/K. Approximated values of the k/K ratio for different joints are listed in Table 17.1.

Examples of the fastener load distributions for multiple-row shear joints are summarized in Table 2.5. According to the analysis, fastener loads are highest at the leading (or lagging) row of holes. Reduced fastener row spacing and a relatively more compliant fastener-hole response, which reduce the k/K-ratio, favor more uniform load distributions and reduce the fastener load at the leading (or lagging) row.

Estimating the Net Section SCF. Panel stresses generated at a particular row of fastener holes by the bearing-mode component of fastener load arise from the superposition of two stress fields. One is produced by the fastener load, Q_i, the other by the by-pass load, Q_{BP}, that is transmitted by the panel across the fastener hole to the next row of fasteners (refer to Chapter 17). The resulting SCF can be approximated by summing estimates of the SCFs associated with the two fields. An estimation procedure and validation are described in Chapter 17. Examples of the variations of the SCFs in different, multiple row shear joints are summarized in Table 2.5. These illustrate three important generalizations:

(i) Increasing the number of rivet-rows beyond two greatly diminishes returns for stress reduction at the leading row. For the lap joints of Table 2.5, an increase in the number of rows from two to three (a 50% increase in the number of fasteners) produces only a 12% decrease in the peak stress.

(ii) With the stresses acting on the leading row 40% to 70% higher than on successive rows, the panel and the fasteners at the leading row are much more vulnerable to failure than the following rows.

(iii) High stresses at the leading row can complicate the task of detecting fatigue damage, NDE, because the lagging row of holes of the front panel face the leading row of the back panel (see Chapter 1.1, Figure 1.2). For multiple-row lap joints with standard heads, the highest SCF is produced at two locations: (i) At the interior face of the leading row of the front panel, and (ii) Directly under the lagging row of the front panel on the far side of the of the panel-panel interface (the interior face of the leading row of the back panel). The detection of fatigue damage at this second location is very difficult when only the front panel is accessible. In that case, the front fastener head, the fastener head-panel interface, the entire front panel and the panel-panel interface are barriers to nondestructive detection.

2.7 BIAXIAL LOADING AND THE EFFECTS OF THE FREE EDGE

The SCFs produced by wide joints with multi-fastener rows display variability along the row. The fasteners immediately adjacent to the free edge display higher SCF-values than those

**TABLE 2.6. THE ELEVATION OF THE STRESS
CONCENTRATION FACTOR, SCF, BY THE FREE EDGES
OF NARROW, L = 30.6 mm SINGLE RIVET-ROW LAP
JOINTS RELATIVE TO INFINITELY WIDE JOINTS
WITHOUT FREE EDGES**

TYPE OF LAP JOINT	PANEL THICKNESS mm	SCF-ELEVATION %	F.E.MODEL
Std. head, no sealant	1.53	6.0	4S-6, 4S-7
	1.00	4.6	4S-10, 4S-11
Std. head, sealed	1.53	7.3	4SA-8, 4SA-9
	1.00	5.1	4SA-12, 4SA-13
Countersunk head, no sealant	1.53	6.7	4C-6, 4C-7
	1.00	5.2	4C-10, 4C-11
Countersunk head, sealed	1.53	7.8	4CA-9, 4CA-12
	1.00	6.8	4CA-12, 4CA-13

located a distance $>3P_1$ from the edge. The findings are summarized in Table 2.6. There appear to several sources of the stress elevations.

Biaxial Loading of Single-Row Joints. Iyer [1,35] has examined the effects of transverse loads on wide, single fastener-row joints. His calculations provide estimates of the SCF for four different biaxial stress states:

(i) $\sigma_2/\sigma_1 = \nu$. This condition is obtained in a wide joint loaded in simple tension without transverse loads because the dimensional changes in the (in-plane) transverse direction are constrained.

(ii) $\sigma_2/\sigma_1 = \nu + 0.5$. This condition is obtained in wide, longitudinal seams of cylindrical pressure vessels

(iii) $\sigma_2/\sigma_1 = \nu + 1$. This condition is obtained in wide, longitudinal seams of spherical pressure vessels.

(iv) $\sigma_2/\sigma_1 \approx 0$. This, the condition for a fastener close to the free edge, was not examined explicitly, but was evaluated with the results for i–iii, above by extrapolation.

Iyer's findings, summarized in Figure 2.12, show that biaxial loading produces modest reductions in the net section SCF of shear joints, For example, the biaxial stress ratios, $\sigma_2/\sigma_1 = 0.25$ and $\sigma_2/\sigma_1 = 0.75$ associated with the uniaxially loaded joint and the cylindrical vessel reduce the SCF by ~0.5% and ~5% respectively relative to the value obtained for $\sigma_2/\sigma_1 = 0$ by extrapolation. In other words, the near-zero biaxial stress ratio associated with fasteners next to the free edge would be expected to elevate the SCF by about 0.5% relative to the fasteners in a wide joint away from the edge.

SCF Elevation in Narrow, Single Row Joints. Calculations summarized in Table 2.6 compare the SCFs of wide (no free edges) lap joints, $P_1 = 30.6$ mm, with those of otherwise identical narrow joints, L = 30.6 mm-wide, single rivet joints with 2 free edges and a setback of $P_1/2$. The proximity of 2 free edges elevates the SCF by from 5% to 8%. The authors believe that the narrow joint *may* simulate the conditions experienced by fasteners in a wide joint

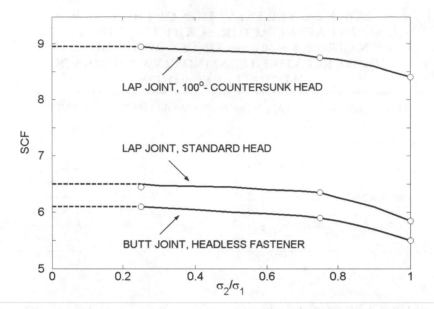

Figure 2.12. The influence of the biaxial stress ratio, σ_2/σ_1, on the stress concentration factors (SCF) of wide, single fastener-row, aluminum butt and lap joint panels with standard head rivets (Models 2P-22, 23 and 24, 3S-16, 17, and 18, and 3C-17, 18, and 19).

adjacent to the free edge. The elevations cannot simply be attributed to the loss of lateral stress at the free edge since the expected increase from this source is only about 0.5% (see preceding paragraph and Figure 2.12). It appears that the SCF elevation at the free edge arises from more subtle changes in the stress distribution. One possibility is that the elevation is related to the set-back—the margin between the fastener and the free edge. Increases of the set-back distance increase the local value of P_1 and are expected to increase the SCF, and vice versa (see Chapter 2.4). It is not clear that a setback of $0.5P_1$ exactly reproduces the stress state obtained in the interior of a wide joint with a pitch of P_1. The effect of setback on the SCF has not been examined here and bears further study.

Biaxial Loading of Multiple-Row Joints. As noted in Chapter 2.6, there are disparities in the amounts of load supported by facing strips of panel between the same fastener rows. The load disparities and accompanying mismatch in the amounts of Poisson's contraction displayed by the facing strips affect the fastener load near the panel edge. Muller [34] has calculated that the mismatch elevates the load applied to the fastener adjacent to the panel free edge by 10%[7]. The elevation is for a 3-row lap joint and $v = 0.33$ and is expected to increase with additional fastener rows and vary directly with Poisson's ratio. It should be noted that

---

[7] This result applies to a 3-row, eight rivets per row, $P_1/D = 5.9$, $D/t = 3.2$ lap joint with a set back of $0.5P1$ at the free edge. The Poisson's ratio of the model is $v = 0.33$.

Muller employed finite element models with a free edge, and therefore including the effects described in the preceding two paragraphs.

2.8 FASTENER STRESSES

Butt Joints. The shear joint panels transmit the load, Q, to the fasteners. Figure 2.13a illustrates the basic loading of the fasteners in butt joints. The edge of the center panel bears against one side of the fastener shank with the load Q, and this is transmitted to the edges of the 2 cover panels on the opposite side. The shear force, Q/2, applied to Section AA produces a moment that is balanced by the shear force acting on Section BB. As the panels bear against the fastener, they compress the shank and generate high levels of contact pressure, whose intensity and angular distribution are described in Chapter 3.4. An expression for the peak tensile stresses, which are produced by bending at location c (Figure 2.13a), has been proposed by Timoshenko [36][8]. A fastener shank with a relatively large diameter-to-length ratio, e.g., such as the aluminum rivet model with $D_S/H = 2$, behaves as a relatively stiff beam that displays little bowing. In that case, the panels transmit essentially no lateral force to the fastener heads, and no net force to the fastener shank. As a result, the head-shank, and nut-shank junction corners (locations 3A and 3B) and thread roots close to the nut, which act as stress concentrators, remain unstressed.

Lap Joint Fasteners, Standard Head. Fasteners of lap joints are subject to four sets of loading shown schematically in Figure 2.13b:

Set 1. The edges of the panels bear against the fastener on opposed sides of the shank with the force Q (locations 1A and 1B). This is analogous to the loading in the butt joint except that the moment of the shear force, Q on the midsection is initially unbalanced. The bearing loads compress the shank under the contacting surfaces.

Set 2. Relative movement between the edges of the panels and shank is resisted by shear tractions, μQ, at locations 1A and 1B arising from friction. These shear tractions contribute to the unbalanced moment and apply tension to the shank.

Set 3. The initially unsatisfied moment produced by the bearing loads at locations 1A and 1B tilt the fastener. The facing panels resists the tilting by generating the normal forces, Q_B, against the underside of the fastener head near the leading corner (location 2A), and the lagging corner (location 2B). These forces produce bending moments and they load the fastener shank in tension. The forces at Locations 2 and 3 are distributed along a ~150° sector of the circumference of the shank and the underside of the rivet head (see Chapter 6). The stress contours for other sections through the fastener are not identical to those in Figure 2.14.

Set 4. Relative movement between the panel and the under side of the fastener head is resisted by shear tractions $μQ_B$, at locations 2A and 2B arising from friction and the normal forces produced by tilting. These shear tractions also produce bending moments acting on the shank.

---

[8] $\sigma^* = 4HQ/\pi D_S^2$

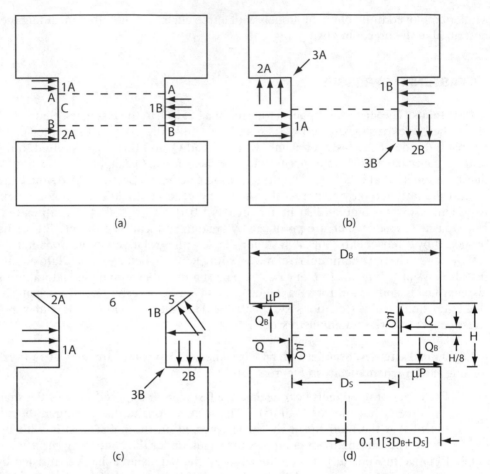

Figure 2.13. Schematic representation of the forces acting on fasteners in joints in the bearing mode: (a) standard head rivet in a butt joint, (b) standard head rivet in a lap joint, (c) 100°-countersunk rivet in a lap joint, and (d) generalized, concentrated force model of the forces acting on standard head fastener in a lap joint. The drawings identify the bearing forces (location 1), the reaction forces (2), and stress concentrations at the shank-head corner (location 3).

A simplified, concentrated force model of the lap joint fastener loading is presented in Figure 2.13d. The locations of the different forces are based on the stress contours in Figure 2.14. The model defines the force, Q_B, as well as the net force, F_{33}, and the average stress, $\sigma_{S,33}$, supported by the fastener midsection:

$$Q_B = Q \, [4 \sin 70°\mu \, (D_S/H) + 1] \, / \, [\sin 70° \, (3D_B/H + D_S/H) - 4\mu] \qquad (2.6)$$
$$F_{33} = Q_B + \mu Q \text{ or} \qquad (2.7)$$
$$F_{33} = (F_{33}/Q) \, Q \qquad (2.8)$$
$$\sigma_{S,33} = 4[(F_{33}/Q)/\pi D_S^2]Q \qquad (2.9)$$

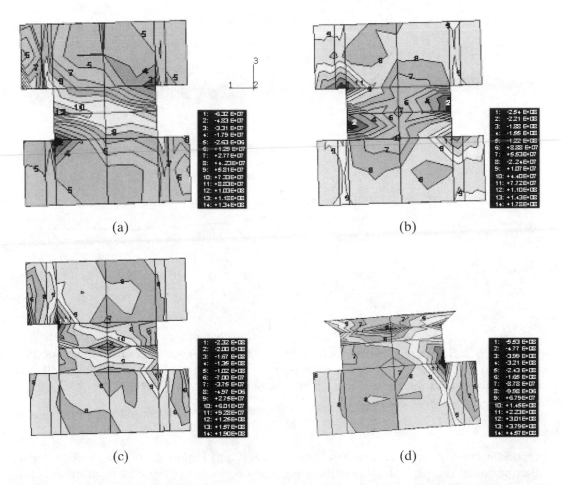

Figure 2.14. Stress contours for the rivets of aluminum, single rivet-row lap joints, $P_1/D = 5$, $D_S/t = 4$, $\sigma = 65$ MPa: (a) σ_{13}-stress contours, standard head, (b) σ_{11}-stress contours, standard head, (c) σ_{33}-stress contours, standard head, and (d) σ_{33}-stress contours, $100°$-countersunk head (Model 4S-8 and 4C-9).

where $D_B \approx 1.55\ D_S$ and (F_{33}/Q) is defined by Figure 2.15. For the rivet model with standard heads, $D_S/H = 2$, and $\sigma = 0.2$, the above relations yield $F_{33} = 0.45\ Q$, and this is close to the value derived from the finite element model in Chapter 6 ($F_{33} = 0.44\ Q$). The authors believe this force model offers useful approximations for fasteners with different shapes.

The four sets of loadings produce complex distributions of stress illustrated in Figure 2.16 which is derived from the stress contours of Figure 2.14a. The shank is loaded eccentrically and the distributions of shear and axial stress are anti-symmetric. High axial stresses produced at the shank periphery near the head-shank corner are ~3x the average tensile stress on the shank midsection. The peak axial stresses are produced by the stress concentrations at two locations: the head-shank junction corner and the roots of threads near the nut-shank

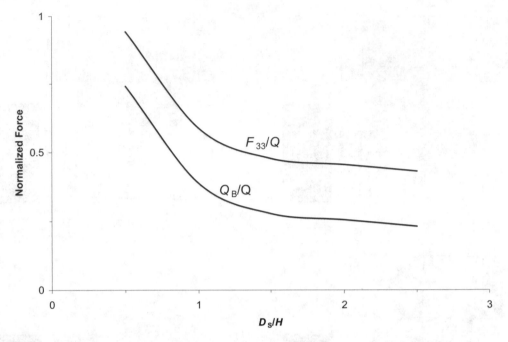

Figure 2.15. Variation of the fastener head axial force-to-fastener load ratio Q_B/Q (see Figure 2.11d), and the average shank midsection force-to-fastener load ratio, F_{33}/Q, with the diameter-to-length ratio D_S/H_1 for lap joint fasteners with standard heads.

junction corner[9]. The present finite element models and Figure 2.16 do not offer reliable descriptions of the peak stresses at these locations because the finite element mesh is too coarse to model the fillet. The following expression for σ_S^*, the peak axial stress, employs the approximation that the peak and average shank midsection stresses are proportional: $\sigma_S^*/\sigma_{S,33} = B$:

$$\sigma_S^* = (4/\pi)B[(F_{33}/Q)/D_S^2]Q \qquad \text{or} \qquad (2.10A)$$
$$\sigma_S^* = (4/\pi)B(F_{33}/Q)(P_1/D_S)(t/D_S)\sigma \qquad (2.10B)$$

The quantity, B, was evaluated by fitting Equation (2.10A) to the fatigue test results ($R = -1, 1 \leq N \leq 2\ 10^8$) reported by Sharp, Nordmark and Menzemer [37] for a 7075-T6, single rivet lap joint ($P_1/D_S = 5.3, t/D_S = 0.34$) fastened with standard head aluminum rivets of

---

[9] For uniform, tensile loading of the shank, these stress concentrations are related to r_F/D_S, the fillet radius- (or root radius-) to-shank diameter ratio[9] [38,39]. Specified head-shank fillet radii for bolts correspond with $r/D_S \approx 0.04$ and an SCF ≈ 4; thread root radii with $5.7 < SCF < 7.5$ [39]. There are two other complications: first, the stressed volumes of these stress concentrations are very small, and secondly, in many cases of interest, local stresses reach and exceed the fastener yield stress. As noted in Appendix B, this tends to reduce the effective SCF.

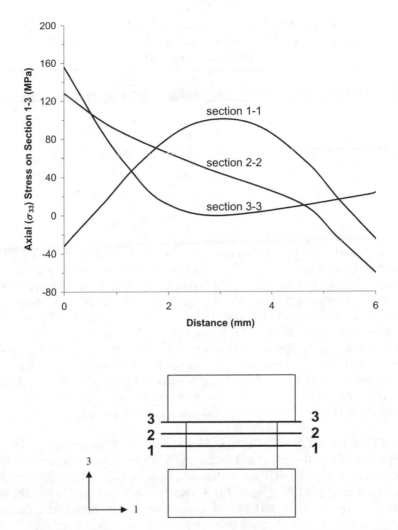

Figure 2.16. The axial, σ_{33}-stress distributions in the shank of a standard head rivet in a wide, single fastener-row lap joint, $P_1/D = 5$, $D_S/t = 4$, $\sigma = 65$ MPa, corresponding with the stress contours in Figure 2.12.

different materials (2024-T3, 2117-T3,6053-T61 and 1100-F). Accordingly, B $\approx$ 2.8 for $1 \leq N \leq 10^3$ and B $\approx$ 4.5 for $10^5 < N \leq 2 \cdot 10^8$. Equation (2.10) is used in Chapter 3.2 and 4.4 to relate the performance of fasteners to their fatigue strength. However, it must be recognized that the generality of Equation (2.10) is uncertain; the quoted values of B are highly approximate and may not apply to steel bolts.

Lap Joints, Countersunk Head. The countersunk head drastically alters the way forces are applied to the fastener and the resulting stresses. The axial component of the force transmitted to the underside of the countersunk head at location 2A in Figure 2.13c is

TABLE 2.7. INFLUENCE OF RIVET HEAD SHAPE ON PANEL SCF FOR SINGLE RIVET-ROW LAP JOINTS ($P_1/D = 5$, $D/t = 4$)

RIVET HEAD	RIVET MATERIAL	TILT ANGLE	SCF	F.E. MODEL
Standard head	aluminum	3.6°	**6.1**	3S-1
100°-countersink to t/2	aluminum	4.2°	**9.8**	3C-1
100°-countersink to t/2	steel	4°	**8.9**	3C-2
100°-countersink to full thickness	aluminum	4.0°	**12.1**	3C-6
61.6°-countersink to full thickness	aluminum	4.5°	**10.5**	3C-5

substantially reduced. This occurs because the inclined surface of the panel pulls away from the mating head when the panel is loaded and because only a component of the bearing pressure is directed axially. At the same time, intense axial forces are generated on the lagging side (location 5) when the wedge-shaped panel presses against the inclined surface of the head. As a result, the tensile stresses on the shank are not symmetrically distributed across the shank midsection but are skewed to the lagging side (Figure 2.14d). The distribution leads to higher peak shank tensile stresses in the countersunk rivet. The in-plane component of the contact forces at location 5 produces a radial compression and a dish-type distortion of the top of the head (location 6). Both of these effects allow the upper panel bending to proceed under the rivet head, and this elevates the panel SCF (refer to Chapter 2.4).

Influence of Fastener on Panel SCF. As already noted in Chapter 2.4, fastener configuration and properties affect the panel net section SCF. Table 2.7 lists results for five different rivet shapes. Compared to the SCF for a standard head, the SCF for a 100°-countersink-to-$1/2$-panel-thickness-rivet is 61% higher. This is obtained because the bulk of the bearing load is applied to the non-inclined part of the countersunk, top panel, the face of the hole between $z = 0.5t$ and $z = t$, accentuating the stress concentration in the panel at $z = t$. In addition, the countersunk head allows more panel bending to proceed under the rivet head. A stiffer but otherwise identical steel rivet reduces the SCF by 10%. Countersinking to full thickness produces a further 23% increase in the SCF even though it is accompanied by slightly less tilting. This may arise from a further skewing of the bearing and tensile stress distributions toward the panel interior face. A 61.6°-countersink-to-full thickness-rivet favors a more uniform distribution of stress in the thickness direction. However, this shape is close to headless, allowing more tilting and bending under the head, and delivers a high SCF. These findings show that effective clamping by the rivet head combined with a uniform loading along the shank serve to reduce the panel SCF.

Stresses in Fasteners. Estimates of the peak cyclic stresses generated in the fasteners of common, riveted aluminum and bolted steel lap and butt joint configurations are listed Table 2.8. The comparison reveals that the peak stress amplitudes for both the aluminum and steel butt joint fasteners and the aluminum lap joint fasteners are below the peak values in the panels; those for the steel lap joint exceed the peak stress in the steel panel. The

TABLE 2.8. ESTIMATES OF THE PEAK STRESS AMPLITUDES PRODUCED BY THE BEARING MODE IN COMMON STEEL AND ALUMINUM BUTT AND LAP JOINT CONFIGURATIONS

JOINT TYPE	Aluminum Butt[1]	Steel Butt[1]	Aluminum Lap	Steel Lap
JOINT AND FASTENER CONFIGURATIONS				
P_1/D_s	5	3.5	5	3.5
D_s, mm	6.12	19.1	6.12	19.1
H, mm	3.06	28.6	3.06	25.4
Fastener Rows	2	2	2	2
PANEL STRESSES				
σ_a, MPa[2]	30	50	30	50
σ_a*, MPa[3]	129	215	129	215
FASTENER LOADS AND STRESSES				
Q_a N[4]	702	21,200	702	21,200
$F_{a,33}$ N[5]	~0	~0	176	--
σ_a*,BENDING, MPa[6]	12	110	--	--
σ_a*,TILTING, MPa[7]	~0	~0	~32	~150
σ_a*,TILT+SCF, MPa[8]	~0	~0	~43	~200
(σ_a*-fastener)/(σ_a*-panel)	0.1	0.5	~0.3	~0.9

[1] Estimates apply to center panels of 1:2:1 butt joints
[2] Panel gross section stress
[3] Peak panel stress amplitude, σ_a*=σ_a (SCF)
[4] For 2-fastener row joint, the fastener load, Q_a=0.5$\sigma_a P_1 t$
[5] Average force on fastener midsection produced by tilting, $F_{a,33}$, derived from Figure 2.13.
For aluminum lap joint (D_s/H)=2, $F_{a,33}$=0.45Q_a and for steel lap joint (D_s/H)=0.75,
$F_{a,33}$=0.68Q_a
[6] Peak bending stress amplitude, σ_a*=$[4HQ_a/\pi D_s^3]$
[7] Stress amplitude produced in shank by tilting, σ_a*=3$\sigma_{a,33}$=3$F_{a,33}$/0.7854D_s^2
[8] Stress amplitude produced by the stress concentration at the head-shank and nut-shank
junctions, σ_a*=1.33$\sigma_{a,33}$*

lap joint fasteners are more highly stressed than butt joint fasteners. The peak stresses in the aluminum rivets are lower than those in the steel bolts. This can be traced to the 67% higher stresses applied to the steel joints and the 3-fold larger D_S/H-ratio of the aluminum rivet configuration, which is practical because of the lighter gages of aluminum panels. The implications for fatigue are examined in Chapter 4.4.

CLAMPING, INTERFERENCE, MICROSLIP, AND SELF-PIERCING RIVETS

3.1 THE CLAMPED, FRICTIONAL OR SLIP-RESISTANT MODE OF BUTT AND LAP JOINTS

Defining the Clamping Force (Bolt Tension). The stress concentrations in a shear joint in the clamping mode depend on both friction and the clamping force or bolt tension, Q_C, (Equation (1.2), Chapter 1.4). Clamping is obtained when the pre-assembly shank height[1], H_0, is less than $2t_0$, the combined initial thickness of the two (or three) panels. The shank must then stretch during assembly to accommodate the misfit. The clamping force, Q_C, is the elastic restoring force, which is related to ε, the elastic strain of the shank after assembly:

$$Q_C = \varepsilon \, E \, A_S \qquad (3.1)$$

Where $\varepsilon = (H - H_0)/H_0$, H is the shank length after assembly, E is the Young's Modulus, and A_S is the cross sectional area of the shank. The average shank tensile stress is[2]:

$$\sigma_S = \varepsilon \, E = Q_C/A_S \qquad (3.2)$$

While the key variable for clamping is the shank strain, ε, it is convenient to use the nominal clamping, %CL, which is easy to evaluate, as a measure of ε:

$$\% CL = 100 \, (2t_0 - H_0)/H_0 \qquad (3.3)$$

---

[1] The distance between the rivet heads or between the bolt head and retaining nut

[2] The tensile stress generated on the shank midsection of a clamped fastener is ~50% higher at the shank surface than in the center.

But %CL, the percent misfit *before* assembly and %ε, the percent misfit after assembly, are not identical. The shank strain is smaller than the nominal clamping because the shank extension is one of three accommodations to the preassembly misfit:

i. The extension of the fastener shank: $\Delta H = (H - H_0) = \varepsilon\ H_0$
ii. The compression of the panels by the fastener heads: $2\Delta t$
iii. Local deformation of the fastener hole edge and the fastener head at the head-shank junction: ΔH_L. The deformations arise from intense contact pressure at the fastener hole edge is illustrated in Figure 3.1(a).

The percent shank strain is the nominal clamping *reduced* by the panel compression and the local deformation.

$$\%\varepsilon = \%CL - 100\ (2\ \Delta t/H_0 + 2\Delta H_L/H_0) \qquad (3.4)$$

The task of evaluating %ε is difficult and has been approached empirically. For example, Bickford [40] has proposed an expression for the $\varepsilon H_0/2\Delta t$ – ratio[3], but without the value of ΔH_L, neither the relation between %ε and %CL nor the clamping force are defined. Chesson and Munse [41] Sterling et al. [42] and Christopher, Kulak and Fisher [43] have dealt with this problem by "calibrating" steel bolts. They measured the strain in the shank for given amounts of preassembly misfit related to the amount of bolt tightening (number of nut-turns). The calibration defines both %ε and %CL, but in the absence of values for the panel compression and the local deformation, the findings are difficult to generalize. The authors have devised an alternative treatment described in Chapter 18 that models the fastener and panels as linear springs. The analysis defines the shank strain and the clamping force for given amounts of nominal clamping. It also provides estimates of the relative contributions of shank strain, panel compression and local deformation. Results obtained for different joints are summarized in Table 3.1. These show that the effective clamping is about 30% to 45% of the nominal clamping.

Clamping Residual Stress Field. Clamped panels are host to a residual stress field that interacts with the stresses introduced when the joint is loaded. For comparable ε, H_0 and t_0, the residual stress fields in butt and lap joint panels are similar. This residual stress field is axisymmetric and displays a prominent, annular region of intense normal compression under the fastener head (Figure 3.1a) as well as circumferential and radial in-plane compression at the outer edge of the hole, and an annular ring of circumferential and radial in-plane tension that peaks at radius, $r \approx 2.0\ r_H$ (Figure 3.1b and c). The field varies in the thickness direction; it is more intense at the top of the cover panel than at the top of the center panel. The examples of the σ_1- and σ_2-residual stresses obtained on the top surface of the cover panel in Figure 3.1 correspond with the level of clamping (%CL = 0.85, %ε = 0.41) employed in steel joints. When the joint is loaded the circumferential residual stresses contribute to the σ_{11}-stress concentration located at $r \approx 2.0 r_H$ as shown in Figure 3.2a and 3.2b. For this reason, the SCFs associated with the clamping mode increase with the amount of clamping.

---

[3] $2\Delta t/\varepsilon H_0 = A_{SH}/A_C$ and $A_C = \pi/4\ [(D_B + 0.1\ H)^2 - D_H{}^2]$ where A_C is the effective area of contact under the bolt head and D_B and D_H are the diameters of the bolt head and the fastener hole, respectively

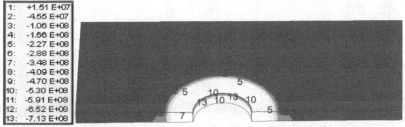

Minimum principal stress, σ_3, provides a sense of contact pressure due to clamping.

(a)

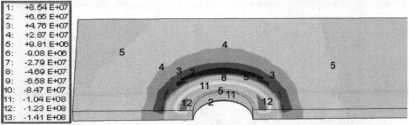

Maximum principal stress, σ_1, gives a sense of $\sigma_{\theta\theta}$ contours. σ_1 is identical to $\sigma_{\theta\theta}$ for $r \geq 1.4\ r_H$ at $\theta = 90°$.

(b)

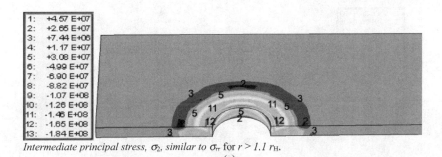

Intermediate principal stress, σ_2, similar to σ_{rr} for $r > 1.1\ r_H$.

(c)

Figure 3.1. Principal stress contours of the cover panel residual stress field produced by the %Cl = 0.85 clamping of a wide, steel, 1:2:1 butt joint, $P_1/D = 5$, $D_S/t = 4$, with standard head fasteners: (a) σ_3, (b) σ_1 and (c) σ_2 (Model 3S-3).

TABLE 3.1. DISTRIBUTION OF ACCOMMODATIONS TO THE INITIAL CLAMPING MISFIT

MATERIAL	JOINT			CLAMPING		SHANK STRAIN[2]	DEFORMATION DISTRIBUTION (%)					MODEL
	H_0	D_S	D_B	%CL	Misfit		Shank			Panels	Local	
	(mm)	(mm)	(mm)		(mm)	%ε	Anal.[1]	FEA[2]	Exp.[3]	Anal.[1]	Anal.[1]	
Aluminum	3.06	6.12	9.79	0.5	0.0153	0.15	31	34	--	20	49	4S-30
Aluminum	6.12	6.12	9.79	0.5	0.0306	0.24	41	49	--	26	33	3S-3
Steel	6.12	6.12	9.79	0.85	0.052	0.41	41	48	--	26	33	3S-4
Steel[3]	108	22	33	1.3	1.4	0.35	44	--	39	35	21	--

[1] Anal.- Analyses and results in Chapter 18 and Table 18.1

[2] FEA - Finite element analyses

[3] Exp. - Experimental determination for a A354BD bolt (43)

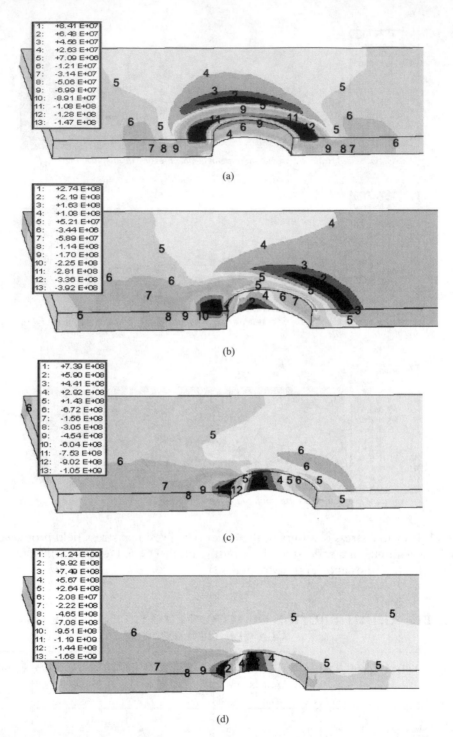

Figure 3.2. The σ_{11}-stress distributions for a wide, single rivet-row, steel butt joint panel in the clamping mode ($P_1/D_S = 5$, $D_S/t = 4$, %CL = 0.85, %ε = 0.41 and μ = 0.2): (a) Residual stresses ($\sigma = 0$), (b) Clamping mode stress field for $\sigma = 126$ MPa, and (c) Post transition stress field for $\sigma = 200$ MPa with relatively weak clamping mode and strong bearing mode components and (d) Post transition field for $\sigma = 260$ MPa with dominant bearing mode component (Model 3S-4).

44

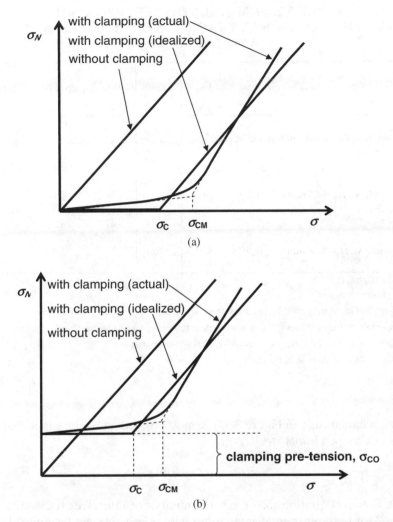

Figure 3.3. Schematic representation of the variation of net section stress, σ_N, with applied stress, σ, showing the effect of clamping: (a) butt joint, (b) lap joint. The friction attending the clamped mode can support an applied stress, $\sigma = \sigma_C$. When σ_C is exceeded, the panels of "idealized" joints will macro slip into the bearing mode, alleviating the net section stress. In actual joints, microslip produces small elevations of the net section stress when $\sigma < \sigma_C$, microslip, micro tilt, and local deformation postpone the transition to macro slip to $\sigma = \sigma_{MC}$.

Stress Carrying Capacity of Friction Tractions. When the clamped joint is loaded, interfacial tractions are generated along the clamped interfaces and the applied load is transmitted from panel-to-panel and fastener-head-to-panel by friction. Initially, to a first approximation, no load is transferred to the fastener shank. Increases in net section stress arising from shank-panel contact are absent; the net section stress remains fixed at the residual stress

TABLE 3.2. NOMINAL STRESS SUPPORTED
BY THE FRICTIONAL TRACTIONS OF CLAMPED BUTT
AND LAP JOINTS

JOINT TYPE	ε^1	$Q_C{}^2$	$\sigma_C{}^3$	$\sigma_{CM}{}^4$	σ_{CM}/σ_C	$f_{BM}{}^5$
		(kN)	(MPa)	(MPa)		%
Butt joint, steel, H_0=6.12 mm	0.0041	25.12				
Center Panel			107	125	1.2	28
Cover Panel				300	2.8	28
Butt joint, Al, H_0=6.12 mm	0.0024	5.00				
Center Panel			21	27	1.3	27
Cover Panel				67	3.1	22
Lap joint, Al, H_0= 3.06 mm	0.0018	3.53	30	57	1.9	25

1. ε-shank strain

2. Q_C-fastener tension

3. σ_C-applied stress supported by frictional tractions of clamping mode (Equation 3.5)

4. σ_{CM}-applied stress corresponding with transition from clamping to bearing mode

5. f_{BM}-fraction of applied stress supported by bearing mode when $\sigma=\sigma_C$;

   %f_{RM}=100(slope of σ_N-σ curve)/SCF$_N$

level, as shown schematically in Figure 3.3. The most stress that can be supported by the friction tractions of clamped joints, σ_C, is[4]:

$$\sigma_C = \varphi\, \mu\, Q_C/(t\, P_1) \qquad\qquad (3.5)$$

where $Q_C = \varepsilon\, E\, A_{SH}$ (Equation (3.1)), φ is the number of interfaces transmitting frictional tractions[5]: $\varphi = 2$ for butt joints and for lap joints with relatively rigid fasteners. The values of σ_C that can nominally be supported by clamping are listed in Table 3.2 for the joints and conditions evaluated here,

Clamped Mode-to-Bearing Mode Transition. When the applied stress exceeds the value of σ_C, the panels are expected to slide (undergo macroslip), ending the clamped-mode or slip-resistant phase. Further increases in applied stress would then be transmitted by the sliding panels to the fastener shank in the bearing mode accompanied by the intensification of the

---

[4] Assuming idealized Coulomb friction.

[5] These are the interfaces that allow the joint to slip into the bearing mode. In lap joints with relatively rigid fasteners, two interfaces, the panel-panel interface and the interface between the underside of the fastener head and the panel, must slip ($\varphi = 2$). In lap joints with relatively compliant fasteners, slip at the fastener head-panel interface is obviated by the shearing of the fasteners ($\varphi \approx 1$).

net section stresses and SCF. The transition is expected to be marked by a distinct change in slope of the σ_N vs σ curves from zero slope to a slope equal to the SCF, as shown schematically in Figure 3.3.

The transitions displayed by the finite element models of the clamped joints, illustrated in Figure 3.4, depart from the idealized picture in three ways:

1. The net section peak stress is not fixed in the clamped state but increases slowly with applied stress.
2. The transition is not abrupt.
3. The stress corresponding to the transition, σ_{CM}, is higher, and in some cases close to 3x the value of σ_C.

The origins of these departures are identified in the next section.

Microslip, Micro-Tilt and Local Deformation. As load is transferred across the panel-panel interface by the frictional shear tractions, stress is gradually transmitted to portions of panels under the fastener head and around the shank. The stressed parts of the panels strain elastically, producing small relative displacements or microslip, which bring the panels into bearing contact with the fastener shank. As a result, a slowly growing portion of the fastener load is transmitted in the bearing mode *while the joint is fully clamped*. The calculations indicate that ~26% of the applied load is transmitted by the bearing mode in the fully clamped condition[6]. This is accompanied by the formation of a corresponding "net section" SCF and the increase of σ_N with applied stress while the joint is in the clamped state as illustrated in Figures 3.2b, and 3.4a and 3.4c. The formation of a net section SCF in the clamped state is expected for joints with no clearance between the fastener shank and panel hole as is the case for riveted connections. Bolted connections with clearance can accommodate microslip without panel-shank contact, and this is expected to delay bearing contact and the development of the net section SCF until σ_C is exceeded.

Microslip, by transmitting part of the load to the shank without reducing the capacity of the frictional tractions, adds to the stress the joint can support in the clamping mode. This elevates σ_{CM}, the stress at the clamped-to-bearing mode transition, by $\approx 26\%$. Two other effects, "microtilt" and the local deformation at the fastener head- panel hole contact, also elevate the transition stress. Frictional tractions applied by the lap joint panels to the underside of the fastener heads tilt the fastener small amounts. The "microtilt" elevates the contact pressure at the corner of the fastener head thereby increasing the frictional tractions. As noted earlier in this chapter and illustrated in Figure 3.1a, clamping produces intense pressure at the fastener head-panel hole edge contact. The pressure deforms the hole edge and displaces the underside of the head by a small amount into panel hole. The authors believe the resulting interlocking of the fastener and the panel interferes with onset of the bearing mode. The interlocking is limited to panels in contact with the fasteners, including both lap joint panels but the cover panels of the butt joint, but not the butt joint center panel. Consistent with this, Table 3.2 illustrates that the elevation of the transition stress, σ_{CM}/σ_C, is small for the butt joint center panel, large for the butt joint cover panel and lap joint panels, and increases with the amount of clamping.

---

[6] The percent of the applied load supported by the bearing mode, %f_{BM}, is the slope of the $\sigma_N - \sigma$ curve in the clamped state divided by the bearing mode net section SCF: %$f_{BM} = 100$ slope/SCF_N

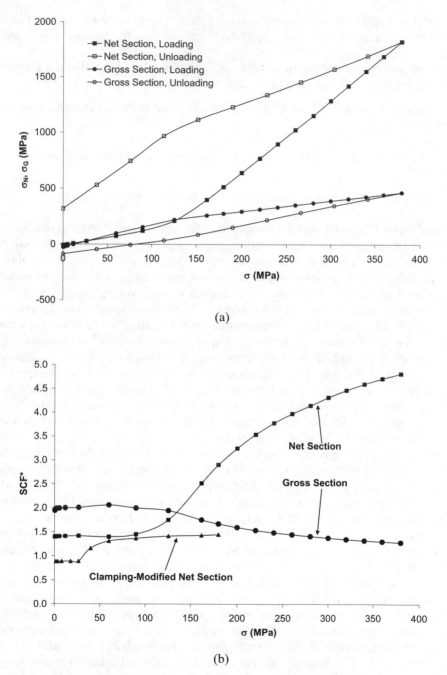

(a)

(b)

Figure 3.4. Variations with applied stress of the peak net section stress, gross section stress and corresponding SCFs for a clamped (%CL = 0.85, %ε = 0.41), wide, single fastener row, steel butt joint panel, P_1/D = 5, D_S/t = 4, (a) and (b) center panel. *(figure continues)*

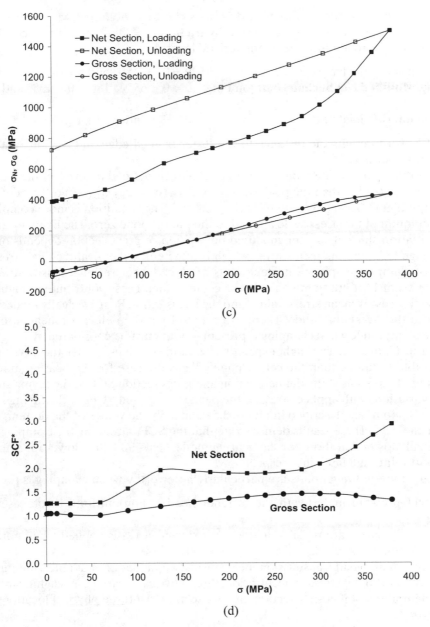

Figure 3.4. (continued) (c) and (d) cover panel (Model 3S-4), [22].

Clamping Mode Stress Field. When the clamped joint is loaded, additional stresses are superimposed on the clamping residual stresses and this produces a distinct clamping mode σ_{11}-stress fields in the panels. With increasing applied load, this stress concentration associated with the residual stress field intensifies and rotates toward $\theta = 90°$ (the loading direction) and a second concentration emerges at $\theta = 0°$ & $180°$ and $r = r_H$. Both of these features are visible in Figure 3.2b which describes the field obtained when the fastener load reaches the value that can be supported by friction $Q = Q_C$. Under these conditions, elevated stresses are found in three characteristic locations:

1. **Net Section** (butt and lap joints): $\theta = 0°$ & $180°$ and $r = r_H$.
2. **Clamping-Modified Net Section** (butt joints): $0 < \theta < 65°$ & $180 < \theta < 135°$ and
                                                         $r \approx 2.0\ r_H$.
3. **Gross Section** (lap joints): $\theta = 90°$ and $r \approx 1.5\ r_H$.

where $r \approx 1.5\ r_H$ represents the distance to the fastener head edge and $r \approx 2.0\ r_H$, a radial beyond the fastener head.

The SCF appropriate for the fatigue of clamped joints is defined in terms of the elevation of the stress range (rather than the peak stress): $SCF^* = (\sigma^* - \sigma_{RESIDUAL})/\sigma$. This definition is employed in Figure 3.4 and 3.5 and in Tables 3.2—3.5. The definition is more complicated when the minimum of the stress cycle applied to the joint is none zero. The likely location of the crack initiation site: net section, modified net section or gross section depends on both the stress range SCF and the mean stress which is affected by the local residual stress. The high levels of clamping employed in steel joints produce higher residual tensile stresses in the cover panels of 1:2:1 butt joints than in the center panel. These contribute to higher values of the peak stresses, mean stress-values and the stress ratios, R[7] in cyclically loaded joints but do reduce the stress amplitude. There is a tradeoff between elevated mean stress and reduced stress amplitude when clamping is present—both must be considered in design. This is illustrated in Figure 3.3. The higher mean stress- and R-values render the cover panels more vulnerable to failure than the center panels. As is the case for the bearing mode, the panel bending of lap joints shifts the net section and gross section SCF to the faying surfaces $(z = t)$. The variations with applied stress of the peak stresses and SCFs at the faying surface of a clamped lap joint are described in Figure 3.5a and 3.5b; the values of the dominant SCF are listed in Table 3.3. These results demonstrate that the SCFs attending the clamping mode are substantially lower that those for the bearing mode: 66%- and 40%-lower for single fastener row butt joints and lap joints, respectively.

In summary, the following points are particularly noteworthy when clamping is present:

1. The contact pressure underneath the fastener head is non-uniform. It increases with applied stress.
2. Microslip and bearing are possible as soon as external stress is applied when non-zero-clearance fasteners are used.
3. In terms of primary load transmission mechanism, two modes are possible: clamping and bearing. But in terms of stress concentration, three physical locations are significant: gross section, clamping-modified-net-section and net section. <u>All three physical locations must be monitored</u>.
4. The transition from the clamping to bearing mode with increasing applied stress is difficult to relate to the onset of macroslip or sliding in the joint.

Combined Clamping-Plus-Bearing Mode Regime. Beyond σ_{CM} only the bearing field intensifies. As the bearing field becomes a progressively larger component of the combined

---

[7] The stress ratio $R = \sigma_{MIN}/\sigma_{MAX}$

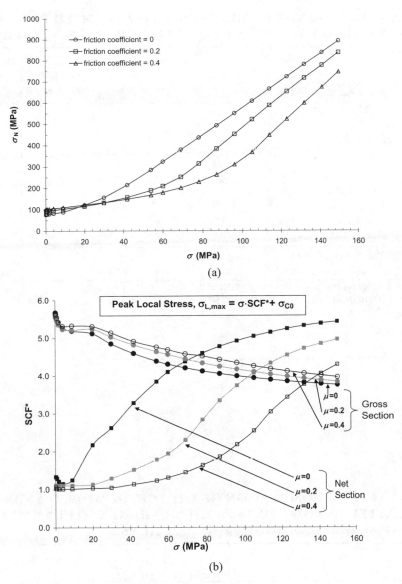

Figure 3.5. Variations with applied stress of the peak net section stress, and net and gross section SCFs for a clamped (%CL = 0.5, %ε = 0.18), wide, single fastener row, aluminum lap joint panel, $P_1/D = 5$, $D_S/t = 4$, for different values of the friction coefficient: (a) net section stress, and (b) net and gross section SCF (Model 4S-29), [22].

field, the net section SCF increases, gradually approaching the value obtained in the absence of clamping. An example of the *"combined"* clamping and bearing field is illustrated in Figure 3.2c. It displays a dominant net section and a weak clamping-modified net section stress concentration. The corresponding variations of the peak stresses and the SCFs are illustrated in Figure 3.3. For the butt joint center panel, the fraction of the applied stress

TABLE 3.3. VALUES OF THE DOMINANT SCF FOR CLAMPED BUTT AND LAP JOINTS, AND LAP JOINTS WITH SELF-PIERCING FASTENERS

	SCF			C (mm/GN)		MODEL
	$\sigma/\sigma_C=1$	$\sigma/\sigma_C=2$	$\sigma/\sigma_C=3$	$\sigma/\sigma_C=2$	$\sigma/\sigma_C=3$	
BUTT JOINT, CENTER PANEL						
$P_1/D=5$, $P_1/D_B=3$, $D/t=2$						
Single fastener-row, steel, $\sigma_c=107$ MPa	1.5^{1}, 2.2^{2}, 1.7^{3}	3.4^{1}	4.4^{1}			3S-2
Two fastener-rows, steel, $\sigma_c=107$ MPa	1.8^{1}, 2.0^{2}	1.7^{1}	2.4^{1}			3S-4
Single fastener-row, aluminum, $\sigma_c=21$ MPa	1.9^{1}	3.4^{1}	4.4^{1}			3S-3
BUTT JOINT, COVER PANEL						
$P_1/D=5$, $P_1/D_B=3$, $D/t=2$						
Single fastener-row, steel, $\sigma_c=107$ MPa	5.5^{1}	3.6^{1}	3.4^{1}			3S-2
Single fastener-row, aluminum, $\sigma_c=21$ MPa	5.5^{1}	3.5^{1}	5.9^{1}			3S-3
LAP JOINT, $P_1/D=5$, $P_1/D_B=3$, $D/t=4$						
Single fastener-row, aluminum, $\sigma_c=15$ MPa	3.6^{3}	4.4^{3}	4.3^{1}			4S-30
WIDE SINGLE FASTENER ROW, 5754-0 ALUMINUM LAP JOINTS WITH CARBON STEEL, SELF PIERCING RIVETS, $\sigma=55.6$ MPa (44)						
$P_1/D_S=15$, $D/t=3$	13.7^{3}	--	--	--	--	--
$P_1/D_S=9$, $D/t=2.5$	12.2^{3}	--	--	--	--	--
$P_1/D_S=9$, $D/t=1.67$	16.7^{3}	--	--	--	--	--

[1] Net section

[2] Clamping modified net section

[3] Gross section

TABLE 3.4. COMPARISON OF THE SCFS OF WIDE, 1- AND 2-FASTENER-ROW JOINTS IN THE BEARING AND CLAMPING MODES; $P_1/D = 5$, $P_2/D = 5$

Joint Type	SCF			SCF_{BP}	FE MODEL
	1-row joint	2-row joint	% change		
BEARING MODE					
Butt joint, Al center panel/cover panel	6.3/6.4	4.5	-30^{1}	2.5	3S-1
Lap joint, Al	6.1	4.4	-29	2.5	4S-1, 5S-1
CLAMPING MODE					
Butt joint, steel center panel	1.8	1.8	0	1.8	3S-2,3S-4

1. Estimated using superposition method described in Chapter 17

TABLE 3.5. ESTIMATES OF THE PEAK STRESS AMPLITUDES PRODUCED IN TYPICAL ALUMINUM AND STEEL BUTT AND LAP JOINTS IN THE CLAMPING MODE

JOINT TYPE	Aluminum Butt	Steel Butt	Aluminum Lap	Steel Lap
JOINT AND FASTENER CONFIGURATION				
P_1/D_S	5	3.5	5	3.5
D_S, mm	6.12	19.1	6.12	19.1
II, mm	3.06	23.6	3.06	25.4
Fastener Rows	2	2	2	2
CLAMPING CONDITION				
%Cl	0.50	0.85	0.5	0.85
%ε	0.17	0.41	0.17	0.41
σ_S, MPa	170	849	170	849
PANEL STRESSES ($\sigma_M=0$)				
σ_a, MPa	30	50	30	50
σ_a*, MPa	51	85	90	150
FASTENER LOADS AND STRESSES				
Q_a, N	182	5512	182	5512
$F_{a,33}$, N	~0	0	82	3800
$\sigma_{a,BENDING}$*, MPa	0^b, 3^c	0^b, 29^c	--	--
$\sigma_{a,TILTING}$*, MPa	~0	~0	0^b, 8^c	0^b, 13^c
$\sigma_{a,TILT+SCF}$*, MPa	~0	~0	0^b, 10^c	0^b, 17^c
σ_m*, MPa	~σ_Y	~σ_Y	~σ_Y	>σ_Y
FS	~25	~60	~25	~30

[a] refer to Table 2.8 for calculational details
[b] with clearance
[c] without clearance

transmitted by the bearing mode when $\sigma > \sigma_{CM}$ is approximated by $(1 - \sigma_C/\sigma)$. Accordingly, the value of the net section SCF in the transition region is:

$$SCF \approx SCF_B \, \xi \, (1 - \sigma_C/\sigma) \tag{3.6}$$

where the SCF_B is the stress concentration produced by the bearing mode alone, and the factor $\xi = 1.1$, corrects for stress field overlap. Figure 3.2d illustrates the field obtained when the applied stress is 2x the value of σ_{CM} and the bearing mode dominates.

The variation in stress during cyclic loading (illustrated for the highest applied stress in Figure 3.3b) displays stick-slip hysteresis as a result of the frictional tractions. It appears that hysteresis attending clamping reduces the local stress variation during repeated load cycling. These findings have important implications. They show that the benefit of the clamping mode, the reduced SCF and reduced stress range, persists after the capacity of the clamping mode has been exceeded. The SCFs produced by the clamping mode are expected to display P_1/D_B- and D_B/t-dependencies similar the P_1/D- and D/t-dependencies displayed by the bearing mode (Figures 2.9 and 2.10).

Butt and lap joints display smaller excess compliance values in the clamping mode compared to the bearing mode (Table 12.10). While the enhanced stiffness leads to a less uniform distribution of fastener loads in multi-row joints it does not elevate the SCF at the leading row of holes of a butt joint. The effects of clamping on contact pressure and microslip at the fastener-panel and panel-panel interfaces are described in Chapter 3.3.

Double Row Butt Joints. The SCF-values for the leading row of holes in clamped, single and double fastener row butt joints are presented in Table 3.4 and compared with single and double row lap joints in the bearing mode. Doubling the number of fastener rows of an *unclamped* lap joint the SCF of the by 28%, and a similar reduction is expected for the *unclamped* butt joint. However, the finite element calculations reveal that the same doubling has virtually no effect on the SCF of the center panel of the *clamped* butt joint. This appears to be a consequence of the low value of the SCF of the clamping mode and the nearly identical SCF_{BP}, the corresponding by-pass SCF (see Chapter 17). The estimate of the SCF_{BP} derived from the single and double row results, $SCF_{BP} \approx 1.8$. is smaller than the corresponding value, $SCF_{BP} = 2.5$, for bearing mode joints. This is reasonable since the fastener hole is shielded from the by-pass stress in clamped joints. While clamping offers large reductions in the SCF of the butt joint, the benefits of multiple rows are sacrificed.

3.2 FASTENERS IN THE CLAMPING MODE

The forces supported by fasteners in the fully clamped condition have two components. One arises from the small part of the total fastener load, about 26%, which is transmitted to the shank by the bearing mode *in the absence of clearance*. This produces the same type of shear, bending and tension stress distributions as described for the bearing mode in Chapter 2.6, but with a nearly 4-fold reduction in the peak stresses. For installations with clearance, this bearing mode contribution to the clamped state is eliminated almost entirely. The second component is the axial tension in the shank produced by clamping and described by Equation (3.2.). These stresses are elevated by the stress concentrations at the head- and nut-junction corners and thread roots. For static loading, the two sets of stresses are simply superimposed. For cyclic loading, the bearing mode stresses cycle about the invariant clamping stress which can be viewed as the mean stress. This implies that clamping, without clearance, produces a nearly 4-fold reduction in the amplitude of the cyclic stresses experienced by the shank, and a 100% reduction with clearance. But clamping also introduces high level high mean stresses. Estimates of the fastener stresses in the clamping mode are listed in Table 3.5 and can be compared with stress amplitudes produced by the bearing mode under comparable conditions in Table 2.8. These tables illustrate that both the peak panel and peak fastener stress amplitudes are reduced by clamping. Cyclic stresses in bolts with clearance are expected to be vanishingly small, however, the peak mean stresses at the head-shank and nut shank corners are elevated to levels that approach or exceed the fastener yield strength.

3.3 HOLE EXPANSION, SQUEEZE AND INTERFERENCE

The fatigue strength of a shear joints in the bearing mode is improved by the cold, forced expansion of the panel hole. The benefits are related to the plastic deformation and residual stresses that are introduced by the expansion of the hole. The stresses, in turn, are affected by differences among several methods for producing the expansion.

1. **Mandrel.** An oversize mandrel is forced through the hole before the fastener is installed [45]. The pressure acting on the hole face is released when the mandrel is removed. Elastic deformation is recovered but the portion of the expansion associated with plastic deformation and the attending cold work and residual stresses remain.
2. **Upset and Squeeze.** A large squeeze force is applied to the driven end of a rivet to plastically upset (and expand) the rivet shank as the head is formed [34,48,47]. In this case, elastic expansion of the shank is recovered and part of the pressure acting on the hole face is released when squeeze force is released. The portion of the hole expansion produced by the permanent expansion of the shank and attending stresses in the panel remain. The relations among squeeze force, the upset, and the amount of hole expansion which depend on the rivet material and dimensions, have been measured and calculated by Muller [34] and Szolwinski and Farris [46].
3. **Interference.** A fastener with a shank larger than the hole or a "taper-lock" fastener whose shank expands during installation is inserted and remains in the hole. There is no release of the pressure on the hole face and the expansion and panel stresses produced during installation remain.

A key parameter for the three methods is the initial hole expansion, %X, (obtained before pressure release) defined as:

$$\%X = 100 \ (D - D_0)/D_0 \qquad (3.7)$$

where D and D_0 are the final and initial diameter of the hole. For cases where elastic modulus of the mandrel or the interference fastener is much higher than the panel, the hole expansion corresponds with the interference:

$$\%I = 100 \ (D_S - D_0)/D_0 \qquad (3.8)$$

Where the fastener and panel possess the same modulus, the expansion is reduced by the elastic contraction of the fastener shank. %C[8] [48] and:

$$\%X = 0.5(1 + \mu) \ \% \ I \qquad (3.9)$$

Axial Variation of the Hole Expansion. The hole expansions produced by the upset and interference methods can vary along the axial (thickness) direction of the fastener hole. Variations of the expansion are obtained even when the interference is uniform and the shank and panel moduli are comparable. The large, rigid fastener head restricts the contraction of the adjacent shank relative to that displayed by the shank at the joint midsection. Figure 3.6a illustrates for 1% interference that the calculated expansion of the hole at z = 0, adjacent to the head, is ~50% larger than the value obtained at z = t, the shank midsection. Even larger

---

[8] $\%C = \%X(1 - \nu)/(1 + \nu)$ (48).

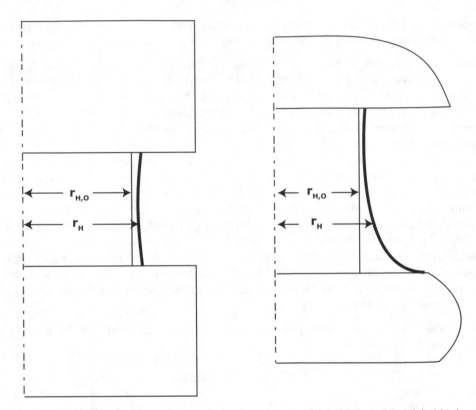

Figure 3.6. Calculated axial variations of the hole expansion produced by (a) 1% interference (4S-20) (1) and (b) 18.9 kN squeeze (46). The hole expansion is magnified 10X.

axial variations have been calculated for the hole expansion produced by upsetting with a large squeeze force [34,46]. For the example illustrated in Figure 3.6b, the hole expansion in the panel on the side of the driven head varies by a factor of five. The possible effects of the axial gradients of hole expansion on the residual stresses are not established and are neglected here. The authors and others [34,46] focus attention on the expansion produced at the panel faying surfaces (at z = t) because this is the location of the stress concentrations produced by the bearing mode.

Residual Stresses Produced by Hole Expansion. A high level of contact pressure must be applied to the periphery to expand the hole. The pressure subjects both the panel and the fastener (or mandrel) to radial compression. The expansion of the hole by elastic deformation is accompanied by circumferential expansion of the panel adjacent to the hole and attending tangential tension. The values of p, σ_r and $\sigma_{\theta\theta}$ (the contact pressure and the radial and tangential stresses) generated at the panel periphery by elastic deformation are related to the amount of hole expansion [48]:

$$p = \sigma_r = -\,\sigma_{\theta\theta} = E\,(\%X)/[100(1 + \mu)] \tag{3.10}$$

where μ is Poisson's ratio[9]. The values of σ_r and $\sigma_{\theta\theta}$ decay exponentially in the panel with radial distance from the hole edge[10].

The stress state in the rim promotes the onset of yielding at the periphery [49] when $P = \sigma_Y/2$, or when:

$$\%X = (\sigma_Y/2E)\,[100(1 + \mu)] \tag{3.11}$$

where σ_Y is the yield stress, or when $\%X = 0.3$ for 2024-T3 panels ($\sigma_Y = 324$ MPa). An annular plastic zone surrounding the hole forms with increased expansion, extending into the panel a radial distance, r_Y. The plastic zone grows quickly, $r_Y/r_0 = 1.3$ at $\%X = 0.4$ to $r_Y/r_0 = 1.6$ at $\%X = 0.8$[11], occupying the bulk of the panel under the rivet head under these conditions.

The elastic-plastic state alters the σ_r- and $\sigma_{\theta\theta}$-stress distributions[12]. The main changes are illustrated in Figure 3.7:

1. The plastic deformation accompanying hole expansion diminishes the tangential tension within the plastic zone. The tangential stress is $\sigma_{\theta\theta} = 0$ at the hole edge when $\%X = 0.6$[13] (or $p = \sigma_Y$) and it is $\sigma_{\theta\theta} < 0$ (compressive) for larger amounts of expansion for when the pressure at the interface is maintained after expansion.
2. For upset-squeeze hole expansion, the pressure at the interface is reduced (but not eliminated) when the squeeze force is released. Figure 3.7b illustrates that this makes the tangential residual stress acting in the plastic zone more compressive.
3. The complete release of contact pressure with Mandrel type hole expansion is expected to produce even higher values of tangential compression and a diminution of the tangential tension away from the hole.
4. The peak tangential tension shifts from the periphery to the plastic zone boundary.
5. For large values of hole expansion, i.e. $0.6 < \%X < 3.0$, the plastic deformation tends to increase the panel thickness. Clamping by the rivet heads resists the growth of panel thickness and leads to the formation of out-of-plane compressive residual stresses in the annular region, $0.2 < r/r_0 < 0.6$ [34]. As a result, the portion of the load applied to the joint that is transmitted by the clamping mode increases with the amount of hole expansion. Calculations by Muller [34] indicate that with $\%X \approx 3$, as much as $\approx 40\%$ of the applied load could be transmitted by the clamping mode[14]. A different model for evaluating the clamping stress produced by squeeze and unloading is treated by Deng and Hutchinson [50].

---

[9] Equations (3.10) and (3.11) apply to 2-dimensional problems, neglect friction and are valid for a large circular, elastic panel with a small hole. when the amount of interference is uniform along the axial direction of the fastener shank and around the circumference [48]. They approximate the conditions in a shear joint panel.

[10] For elastic behavior, $\sigma_{\theta\theta} = -\sigma_r = p(r/r_0)^{-2}$ where r_0 is the hole radius and r the radial distance from the interface [49].

[11] Estimates derived from 2-dimensional finite element analyses for elastic-perfectly plastic material behavior, σ_Y (2024-T3) – 324 MPa, $\mu = 0.2$, and multi-rivet constrains along the panel edges. In the absence of such constraints, the zone grows exponentially [49,50]: $r_Y/r = \exp\,[(p/\sigma_Y) - 0.5]$.

[12] A 2D-analysis of hole expansion without pressure release for elastic-perfectly plastic behavior offers the following relations [49]:

$\sigma_r = -p + \sigma_Y \ln(r/r_0)$ when $r_0 \leq r < r_Y$ and $\sigma_r = -0.5\sigma_Y(r/\psi r_0)^2$ where $\psi = \exp[(p/\sigma_Y) - 0.5]$ when $r > r_Y$, and

$\sigma_{\theta\theta} = (\sigma_Y - p) + \sigma_Y \ln(r/r_0)$ when $r_0 \leq r \leq r_Y$ and $\sigma_{\theta\theta} = -0.5\sigma_Y(r/\psi r_0)^2$ where $\psi = \exp[(p/\sigma_Y) - 0.5]$ when $r > r_Y$

[13] For 2924-T3 Al with $\sigma_Y = 324$ MPa.

[14] The estimates apply to a three rivet row lap joint subjected to a cyclic stress of $\sigma_a = 40$ MPa [34].

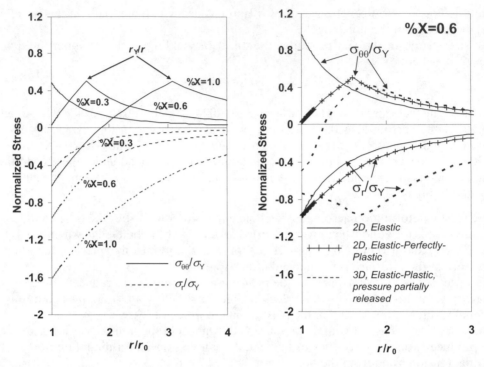

Figure 3.7. The variation with radial distance from the edge of the fastener hole in a 2024-T3 panel (σ_Y = 324 MPa) of the radial and tangential residual stresses, σ_r and $\sigma_{\theta\theta}$, produced by different amounts of hole expansions, %X: (a) the residual stresses produced by 2-dimensional, elastic-perfectly-plastic deformation [49], and (b) the residual stresses produced by 2-dimensional elastic deformation, 2-dimensional elastic-perfectly-plastic deformation [49], and 3-dimensional, elastic-plastic upset followed by the release of the squeeze force [34].

Effects of Hole Expansion on Bearing Mode Stresses. The tangential residual tensile stresses produced by hole expansion can be viewed as a mean stress to which bearing mode cyclic stresses are added when the joint is loaded. The net effect is not a simple addition. For the case where the panel remains elastic, Figure 3.8 shows that while the residual tension increases with hole expansion, the net section SCF (the slope of the σ vs σ_N curve) decreases markedly. Stress contours produced by elastic-plastic behavior when bearing mode stresses are added to the interference stress field are illustrated in Figure 3.9. These also show that hole expansion and attending plasticity completely remove the net section stress concentration at the hole edge. The peak cyclic stress variations are a small fraction of the applied cyclic stress and their location is shifted radially away from the hole to the plastic zone boundary [34]. Large stress variations are replaced by cyclic plastic deformation. The net effect is a dramatic improvement in the fatigue strength of shear joints which is described more fully in Chapter 4.3.

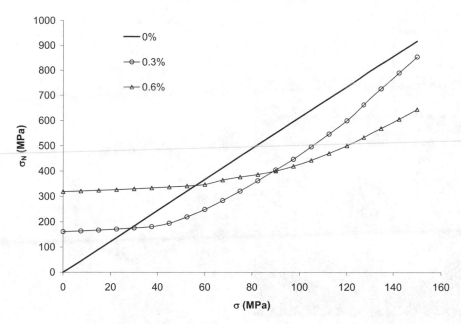

Figure 3.8. The influence of the nominal stress, σ, and 0.3% and 0.6% interference (0.2% and 0.4% hole expansion) on the peak net section, tangential stress, σ_N, for fully elastic behavior of an aluminum butt joint panel. The results are for a 2D analysis of a wide, single row joint, $P_1/D_S = 5$, $D_S/t = 4$ fastened with aluminum pins (Model 3P-40).

There are two additional effects of hole expansion. The attending residual stresses reduce the joint excess compliance. This produces modest increases of the fastener load generated at the leading row of holes. The elevated contact pressure at the shank-hole interface together with friction acts to reduce microslip and fretting. This is described in the next section.

3.4. CONTACT PRESSURE, MICROSLIP AND TANGENTIAL STRESS

Variations of contact pressure, microslip and tangential stress around the periphery of the fastener hole in a butt joint are illustrated in Figure 3.10. These results, obtained in the bearing mode, are derived form elastic-plastic calculations, and illustrate the complexity of the distributions. The bearing mode produces relatively high contact pressures with relatively steep gradients along the circumference. Peak values of microslip (Figure 3.10c) are found at locations with little contact pressure (Figure 3.10a). The peak value of tangential stress (Figure 3.10b) corresponds with the site of the SCF. In lap joints, the variations of contact pressure, microslip and tangential stress are similar but are affected by panel bending. The largest values at the hole periphery are obtained at z = t, adjacent to the interior panel surface, as shown in Figure 3.11. For comparable geometries and loads, the peak values of contact pressure, microslip and tangential stress are generally higher for lap joints than for the

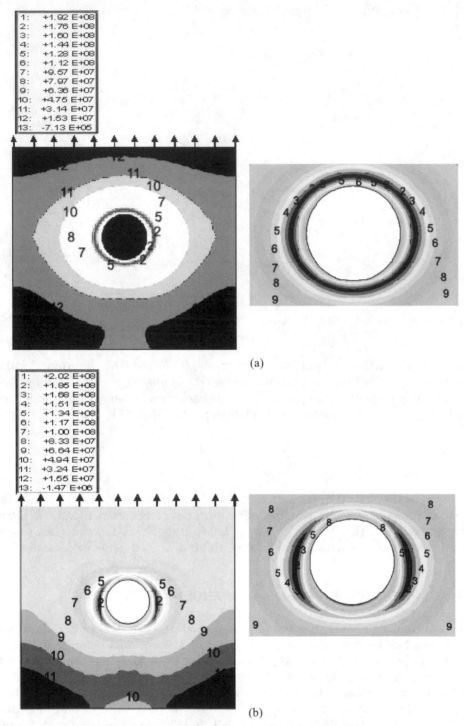

Figure 3.9. The maximum principal stress contours surrounding the fastener hole in a 2-dimensional mode of a wide, single-rivet row, 2024-T3 aluminum butt joint, $P_1/D_S = 5$, $D_S/t = 4$, $\sigma_Y = 324$ MPa for aluminum rivets installed with 0.6% interference (0.4% hole expansion): (a) residual principal stress contours after hole expansion, and (b) principal stresses after hole expanded panel subjected to a nominal stress, $\sigma = 54$ MPa (model 3P-41). Right-hand figures show inner section magnified $\times$ 2.

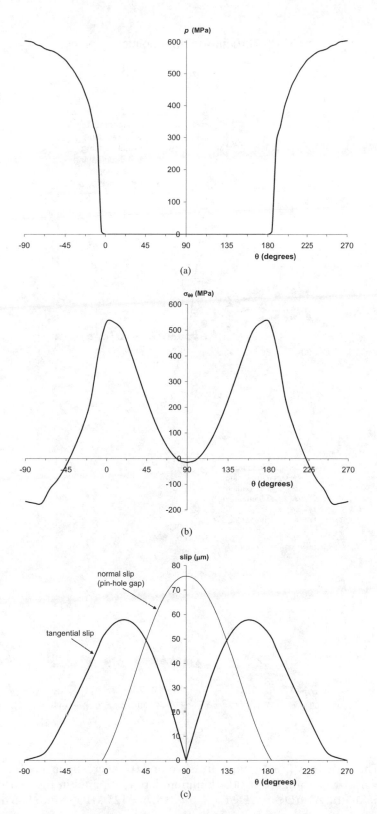

Figure 3.10. The variation of (a) contact pressure, (b) tangential stress and (c) microslip around the periphery of a wide, single fastener-row, pinned butt joint panel ($P_1/D = 5$, $D/t = 4$, Model 3P-25) in the bearing mode for an applied stress, $\sigma = 125$ MPa.

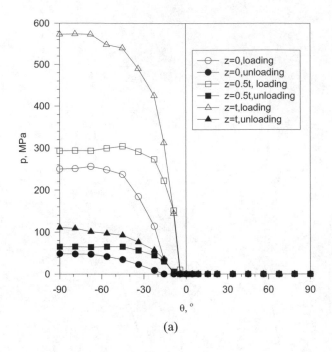

(a)

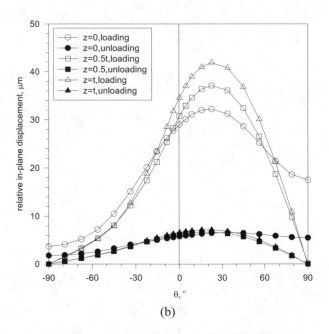

(b)

Figure 3.11. The variation of (a) contact pressure, and (b) microslip around the periphery of a wide, single rivet-row, standard head, aluminum lap joint panel in the bearing mode as a function of the depth, z, $P_1/D_S = 5$, $D_S/t = 4$, $\mu = 0.5$, $\sigma = 125$ MPa (Model 4S-1).

**TABLE 3.6. VALUES OF THE CONTACT PRESSURE,
MICROSLIP AND TANGENTIAL STRESS GENERATED
UNDER THE FASTENER HEAD OF A LAP JOINT IN
THE CLAMPING MODE ($P_1/D_S = 5$, $D_S/t = 4$, $D_B/D_S = 1.6$)**

JOINT TYPE	%Cl	σ, MPa	$\sigma_{\theta\theta}$, MPa	p, MPa	δ, μm
Aluminum Lap Joint	0	30	120	123	10
	0.5	30	131	101	7
Aluminum Butt Joint	0	30	30	18	16
	0.5	30	34	70	7
Steel Butt Joint	0.85	126	80	360	9

butt joint. The influence of clamping on the contact pressure, microslip and tangential tension generated directly under the edge of the fastener head is described in Table 3.6.

Figure 3.12 summarizes the effects of first installing 1% interference (0.65% hole expansion) in an aluminum butt joint panel, then loading the panel to a stress, $\sigma = 125$MPa, and finally unloading. Before hole expansion, contact pressure at the interface and tangential tension in the adjacent panel are absent. Loading the panel produces a peak microslip of $\delta = 60$ μm at the $\theta = 30°$ and 150° locations. The expansion produces high contact pressures and tangential stresses and these are intensified in some locations when the joint is loaded (Figures 3.12a and 3.12c). Friction combined with the high contact pressures lead to a ~5-fold reduction of the peak microslip and the microslip the amounts of microslip (Figure 3.12b). The implications for fretting wear and fatigue are examined in Chapter 4.5

3.5 SELF-PIERCING RIVET

The self-piercing rivet has a countersunk head and a hollow shank. In self-piercing riveting (SPR), a joint is formed by forcing the high-strength tubular rivet through overlapping panels of a softer material using the axisymmetric punch and die arrangement illustrated in Figure 3.13 [44]. The upper sheet is pierced through its entire thickness by the rivet and the lower sheet is pierced only partially. Piercing forces also cause the lower sheet to flow into the die cavity locally and conform to the cavity shape. The piercing operation naturally produces interference, clamping and intimate frictional contact. In addition, the dimple-like, out-of-plane contours of the panels, similar to those seen in joints formed by the clinching process, resist the shearing of the panel and interfere with the transfer of the fastener load to the rivet shank. The in-plane shear tends to produce lateral panel separation, tension in the shank and a clamping action by the rivet. Consequently, the performance of SPR joints has similarities to conventional rivets installed with interference and clamping. Consistent with this, the FEA analyses show that the peak stresses are generated at the $\theta = 90°$ location under the rivet head, essentially the same location found for fully clamped, frictional mode load transfer [44]. The SCFs calculated for three SPR joints are listed in Table 3.3. It should be noted that the calculations for the self-pierced joints account for neither the clamping, interference nor the compressive residual stresses introduced by the piercing operation. For this reason, they probably over-estimate the SCFs of the joints.

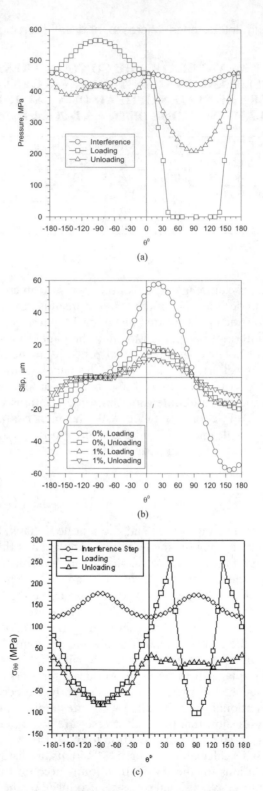

Figure 3.12. Influence of 1% interference on (a) contact pressure, (b) micro-slip and (c) tangential stress for a wide, single rivet-row, aluminum butt joint, $P_1/D_S = 5$, $D_S/t = 4$, $\mu = 0.5$, $\sigma = 125$ MPa (Model 2P-26).

64

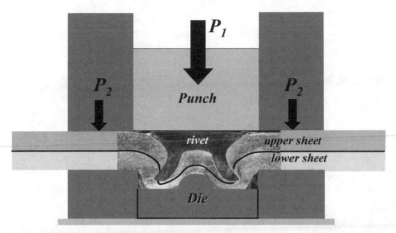

Figure 3.13. Schematic of the punch and die arrangement for installing self piercing rivets. The clamping force, P_2, is applied to prevent lateral movements as the punch forces the rivets through the sheets [44].

Some perspective on the virtues of the self-piercing fastener is obtained from a comparison with conventional fasteners. To facilitate this, the SCFs displayed by the same self-pierced installations in the form of a joint with $P_1/D = 5$ and $D/t = 4$, are estimated with the aid of Figure 2.9. The adjusted values, SCF = 8, 8.9, and 11.1 are comparable to the value, SCF = 8.8, obtained by Iyer (1) for an aluminum lap joint fastened with a conventional, countersunk, steel rivet.

FATIGUE OF SHEAR JOINTS

4.1 STAGES OF THE FATIGUE PROCESS

Fatigue, the loss of load carrying capacity with repeated loading, proceeds in three stages: (i) crack initiation, (ii) stable crack growth and (iii) fracture. In the *crack initiation stage*, the repeated cycles of tensile stress and plastic strain progressively damage the microstructure of the material in isolated locations. The accumulated damage ultimately produces a fully formed, microscopic crack. The per-cycle damage increases either with the local stress amplitude or the mean stress. For this reason cracks initiate at the sites of stress concentration, or sites combining high stresses and intense fretting. Locations of reported crack initiation sites, important for NDE, are identified in Figure 4.1 [34,51-54]. In the second, or *crack growth stage*, the crack extends by small, regular, sub-micron and micron size increments with each repeated loading cycle. The growth trajectory is normal to the direction of cyclic tension. The growth stage ends with the stress cycle that produces a crack length just short of the critical value corresponding with crack instability. The *fracture* stage proceeds during the next and terminal cycle which produces rapid, unstable extension of the crack, and the rupture of the remaining bridges of unbroken material of either the panel or the fastener.

The three stages of the fatigue process are generic for metallic materials and structural components. However, the fatigue failure of structural shear joints can be more complex because it results from mechanical contact at multiple locations between the panels and fasteners and between the panels themselves. The finite element models featured in this text treat the intricacies of mechanical contact. But, the finite element models featured in this text do not contain cracks and so do not model crack growth or fracture. The reader is referred to references [52,55–57] for detailed finite element and fracture mechanics analyses of growth.

4.2 FAILURE MODES AND LOCATIONS

Shear joint panels exhibit three distinct fatigue failure modes important for NDE [34,51–54]; fasteners are vulnerable to two failure modes. These are related to the crack initiation sites.

i. **Net Section Failure.** For joints dominated by the bearing mode without clamping or interference, the crack initiation and stress concentration sites are found near the net

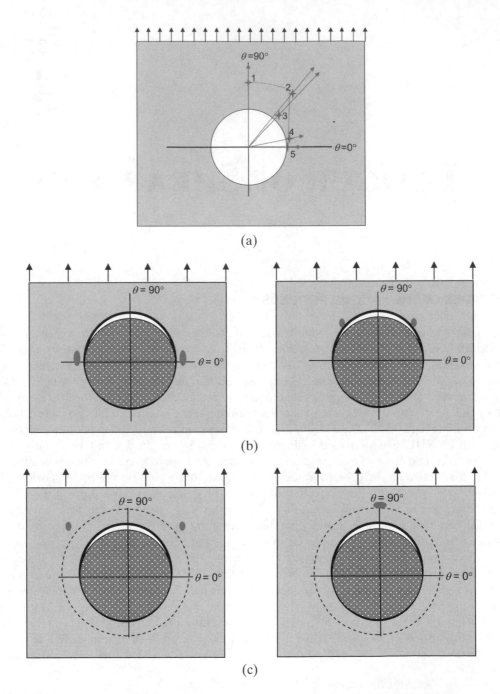

Figure 4.1. (a) Reported fatigue crack initiation sites observed on the faying surfaces of butt and lap joint panel: Locations 1-3 are observed for clamped aluminum and steel joints [51–54], with Locations 1-2 close to the leading edge of the fastener head, Location 4 is observed for aluminum joints with interference [34], and Location 5 is observed for steel and aluminum joints with relatively low levels of interference and clamping [52]. (b) SCF locations computed in butt and lap joints with interference. The SCF moves from the $\theta = 0°$ to $\theta \approx 45°$ with increasing applied stress. (c) SCF locations computed in butt and lap joints with clamping. The boundary of the fastener head is indicated by the dotted line.

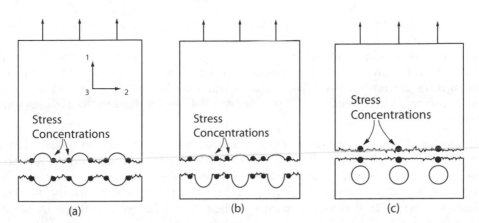

Figure 4.2. Schematic representation of the net section, the clamping- and the hole expansion-modified net section and the gross section modes of panel fracture. The connection between the location of the SCF and the fracture path is illustrated: (a) Net section mode for bearing mode joints, (b) Modified net section modes for joints with clamping and interference, and (c) gross section mode for clamped lap joints with large hole expansions and fretting fatigue failures [34,51–54].

section locations (Figure 4.1a) at the faying surface. These sites may also exhibit fretting wear. Fatigue cracks formed at these locations extend into the panel in the two-direction (normal to the loading direction) and in the three-direction (towards the exposed panel surfaces). They either intersect adjacent fastener holes or merge with cracks initiated at the adjacent holes. This produces the type of panel fracture shown in Figure 4.2a.

ii. **Clamping- and Interference-Modified Net Section Failure.** For butt joints with hole expansion and those in the clamping mode, high stresses are generated away from the hole edge at angular locations typically $20° < \theta < 70°$ (Figure 4.1b). Fatigue cracks formed at these locations extend into the panel the two-direction (normal to the loading direction), and either intersect or just graze adjacent fastener holes or merge with cracks initiated at the adjacent holes. Fretting wear is invariably present in the failure region. This produces the type of panel fracture shown in Figure 4.2b.

iii. **Gross Section Failure.** The dominant stress concentration in lap joints in the clamping mode, as well as the site of intense fretting site for butt and lap joints is located under the fastener head at the gross section-location (Figure 4.1and 4.2c), either directly under the leading edge of the fastener head or at the faying surface. Gross section failures are also produced by joints with large amounts of hole expansion, but these initiate at the faying surfaces ahead of the fastener head ($r \approx 2r_H$) Fatigue cracks initiated there, grow in the two-direction and tend to bypass adjacent fastener holes. This produces the type of panel fracture shown in Figure 4.2c.

iv. **Fastener Head-Shank Corner or Thread Root Failure.** Rivets and bolts harbor stress concentrations at the fastener head-shank junction corner and, for bolts, at the thread roots near the retaining nut-shank transition (Figure 2.13b). The cracks tend to initiate at these locations, grow normal to the shank axis and truncate the fastener. This is the likely mode of failure for lap joint fasteners butt joints fasteners with threads exposed to bending stresses [39].

v. **Shank Failures.** Cracks initiated by the bending of butt joint fasteners will form at the periphery of the shank, half way between the fastener heads (or head and nut), when thread roots are not involved. These cracks will grow normal to the shank axis. This is not a common mode because the panels are likely to fail before the fasteners under these conditions.

The mode of failure is a signature of either the dominant mode of load transfer (bearing vs clamping), hole expansion, or fretting fatigue. For clamped lap joints it is difficult to distinguish between conventional and fretting fatigue because both mechanisms favor the gross section mode. The fatigue lives of joints show a stronger correlation with the SCF, stress amplitude and mean stress rather than the specific mode of failure, and this is examined in Chapters 4.3 and 4.4.

Multiple Site Damage, MSD. The failure of wide shear joints with many fasteners in a row involves a complex evolution in structural response. Cracks do not initiate simultaneously at each fastener hole because of the stochastic nature of the fatigue process and variations in nominally identical fastener installations. The local stiffness of the panel is reduced after one or more cracks emerge from a fastener hole. This changes the distribution of the fastener loads: the load transmitted to the cracked panel hole is reduced while the loads transmitted to adjacent holes (holes not yet cracked) are elevated. As a result, the rate of growth of the local crack slows, and the likelihood of the initiation of cracks at adjacent fasteners increases. This favors a more uniform distribution of fatigue damage, the formation of multiple, cracked fastener holes, referred to as "multiple-site damage" (MSD), [52,58–64]. It also extends the life of the growth stage. It has been proposed that MSD makes it more difficult for stiffeners to arrest unstable cracks extending in shear joints of fuselage structures [58–64]. Similarly, in multiple fastener joints with fatigue vulnerable fasteners, the fracture of the first fastener does not immediately compromise the shear joint. However, the failure will transfer more cyclic load to adjacent fasteners with additional fastener failures and load redistribution until the joint severs.

4.3 FATIGUE STRENGTH OF PANELS

Components of the Fatigue Life. The term, *fatigue strength*, is used here to denote the stress amplitude a component can support for a specified fatigue life, i.e., a given number of stress cycles The fatigue life is the sum of the number of stress cycles attending the initiation and stable growth stages[1]. For convenience, the crack initiation life is arbitrarily defined here as the number of cycles consumed by the nucleation of the crack and its growth to 1 mm-length.[2] The crack growth life is the number of cycles consumed by continued growth—from the 1mm-length to instability and fracture.

---

[1] The life of the fracture stage is 1 stress cycle and is negligible.

[2] Defined in this way, the initiation life includes the nucleation of the crack and the increment of growth affected by the SCF. It also represents, approximately, the life consumed while the crack is small enough to be hidden by the rivet head.

In the recent past much effort has been invested in fracture mechanics methods to predict the growth life of shear connections. This is important because the growth life corresponds with the inspection life, the period when cracks can be detected by NDI, and its definition makes it possible to schedule inspections so cracks will be detected before the joint fails. Prediction methods for the initiation life are complicated by the need to define the local stress amplitudes, i.e., the relevant SCF of the joints.

SCF-FS Analysis of Shear Joints. A number of worker have demonstrated that the S-N curve (stress amplitude-cycles to failure relation) of a notched component is approximated by the S-N curve of the material divided by the SCF of the notch [65,66]. It is the basis for an estimation procedure, which the authors refer to here as the "Stress Concentration Factor-Fatigue Strength (SCF-FS) Analysis". The analysis makes it possible to estimate the fatigue strength of components with stress concentrations when: (a) the S-N curve for the material (unnotched samples) and (b) the relevant SCF of the joint are known, and provided: (c) any loss of fatigue life from fretting, corrosion and other sources are negligible. Since the SCF is a concept derived from linear elastic material behavior, the method is only reliable as long as stresses within the stress concentration do not exceed the yield stress. The basis for the SCF-FS analysis, other limitations and validation of the method are presented in detail in Appendix B.

The SCF-FS analysis was successfully employed by Seliger [32] in 1943 to derive SCF values for riveted aluminum lap joint panels from S-N measurements. In the absence of independently evaluated SCF values for the panels it was not possible to confirm Seliger's findings. The SCF-values reported in Chapters 2 and 3 belatedly validate Seliger's findings (see Appendix B) and can test the reliability of SCF-FS analysis. In the remainder of this sections, predictions of the analysis are compared with fatigue test results for aluminum and steel butt and lap joints in both the bearing, clamping and adhesive modes of load transfer. The comparisons illustrate that the SCF-FS analysis can provide useful strength predictions for shear connections when these are subjected to low, nominal stress amplitudes and display fatigue lives, $N > 10^5 - 10^6$ cycles.

SCF-FS Analysis of Aluminum Lap Joints Panels in the Bearing Mode. Predictions of the of S-N Curves of riveted aluminum lap joints in the bearing mode obtained with the SCF-FS analysis are compared in Figures 4.3 and 4.4 with fatigue test data for the joints collected by the Strength and Analysis Group of EDSU [67]. The tests examined multiple rivet and multi row joints involving different P_1/D-values and stress cycles with different values of means stress. The appropriate SCF values were obtained using the data reported in Chapters 2 and 3 and adjusted for pitch and panel thickness, using Figures 2.9 and 2.10, and number of rows using methods described in Chapter 17. The analyses employ the average material S-N curve for the material derived from the range of results reported in the literature (see Figure B.3). The Gerber diagram (Figures B.?) was used to adjust the material S-N curve for the mean stress applied to the joint. No information about either the amounts of clamping and interference installed in these joints or the negative contribution of fretting fatigue is available. The reduction of the SCF associated with these sources was offset by neglecting fretting fatigue and the ~8% to 10% increase in the SCF for fastener holes adjacent to the free edge.

In spite of the uncertainties introduced using estimates of the material S-N curve, the predictions are in reasonable accord with the measurements for lives, $N > 10^6$. As is the case for

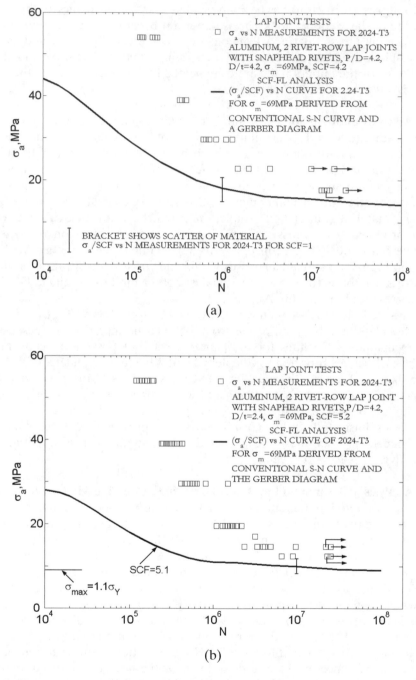

Figure 4.3. Comparisons of fatigue test data for 2024-T3 aluminum, 12 rivets, 2 rivet-row lap joints fastened with snaphead rivets [30] with the predictions of the SCF-FS analysis: (a) D/t = 4 and (b) D/t = 2.4. In each case, the test results are represented by discrete data points. The solid line is the prediction of the SCF-FS analysis (Model 5S-1), [67–70].

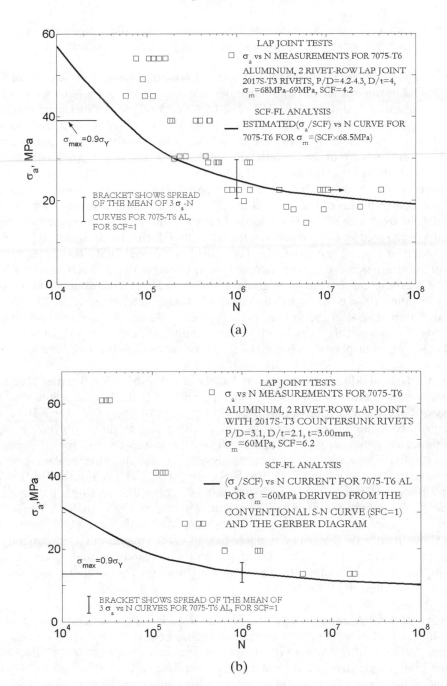

Figure 4.4. Comparisons of fatigue test data for 7075-T6 aluminum, 12 rivets, 2-rivet-row lap joints [30] with the predictions of the SCF-FS analysis: (a) joint with snaphead rivets, (b) joint with countersunk rivets. In each case the test results are represented by discrete data points. The solid line is the prediction of the SCF-FS analysis (Models 5S-1 and 5C-1), [67–70].

the open hole panels (Figure B.1), the predictions understate the fatigue strength with a growing divergence for shorter lives, $N < 10^6$, when the peak stress exceeds the yield stress.

SCF-FS Analysis of Steel Butt Joint Panels in the Clamping Mode. Predictions for bolted and riveted steel butt joints in the clamping mode are compared with S-N data for the joint materials in Figures 4.5 and 4.6 [53,54,71–75]. Included are joints fabricated from three grades of steel, A514 ($\sigma_Y = 827$ MPa), A440 ($\sigma_Y = 349$ MPa) and A7 ($\sigma_Y = 235$ MPa). Material S-N data for the A514 steel is from the same plates used to fabricate the joints. The S-N data for the other steels are taken from the literature and do not correspond with the actual joints. All of the tests were performed on 1:1.6:1 butt joints, a design which assured that the fatigue failures would be confined to the center panels because the nominal stresses supported by the center panels of these joints are 25% higher than the stresses applied to the cover panels. The estimates of the SCFs derived for the joints are based on the reported initial bolt tension reduced by 10% to account for relaxation during the fatigue test [74,76]. The test results for the riveted butt joint display considerable variability consistent with the observation that in this case the rivet clamping force varied between 5 KN and 89 KN [74].

The SCF-FS predictions for the three grades of steel grade agree reasonably with the joint test results in range $10^5 < N < 2 \times 10^6$. Calculations summarized in Table 3.3 illustrate that SCFs in the clamped state at the net section, clamping modified net section and gross section locations are comparable. Consistent with this finding, the center panels of the clamped A514 and A440 joints displayed examples of the three types of failures (Figure 4.1).

SCF-FS Analysis of Aluminum Lap joint Panels in Adhesive-Plus-Bearing Mode. The SCF-FS analysis is useful for estimating the influence of sealants and adhesives on the fatigue lives of shear connections [4–8]. This is illustrated in Figure 4.7 for 3-rivet, single rivet-row, 7075-T6 aluminum lap joints assembled with a low-modulus, polymer sealant at the panel-panel and fastener-panel interfaces. The SCF-values for the sealed joints were obtained with the Thin Adhesive Layer Analysis (TALA) and the sealant properties described in Chapters 14 and 15 [6,21]. The S-N curve of 7075-T6 is adjusted for the mean stress and different SCF-values. The SCF-reduced S-N curve intersecting a particular joint test results (σ_a, N) defines the SCF-value attributed by the SCF-FS Analysis. The numbers next to the data point are the SCFs of the joints calculated with the TALA. The results in Figure 4.7 show that the two sets of SCFs agree to a good approximation. Consequently, both the fatigue tests and the TALA/SCF-FS analysis show a low modulus sealant can produce a ~10-fold improvement in the fatigue life. Analogous findings for bonded steel joints have been reported by Albrecht and Sahli [77].

SRF-FS Analysis of Aluminum Lap Joints with Hole Expansion. The residual stresses and local plasticity introduced by hole expansion and described in Chapter 3.3 produce substantial improvements of the fatigue strength of shear joints. Findings for the effect of hole expansion on the fatigue strength of aluminum lugs are summarized in Table 4.1 [78–80]. The lugs are loaded like the center panel of a butt joint with hole expansion produced with steel pins or bushings installed with interference. The conditions examined include small amounts

---

[3] 1:1.6:1 are the relative thickness valus of the butt joint cover-center-cover panels.

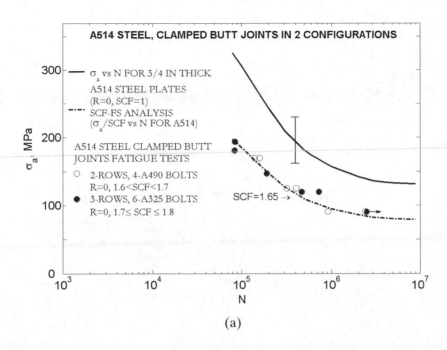

(a)

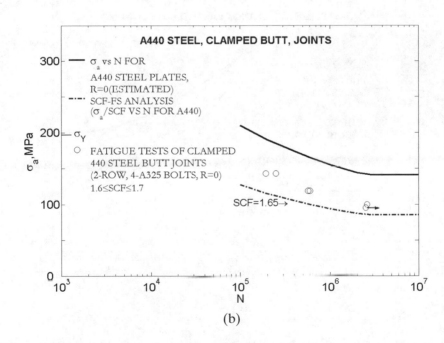

(b)

Figure 4.5. Comparison of fatigue test data for clamped, steel butt joints with the predictions of the SCF-FS analysis (a) A514 steel butt joints with A490 bolts [52], and (b) A440 steel butt joints with A325 bolts [53,54,71–73].

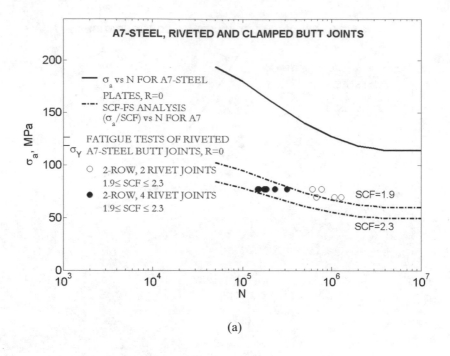

(a)

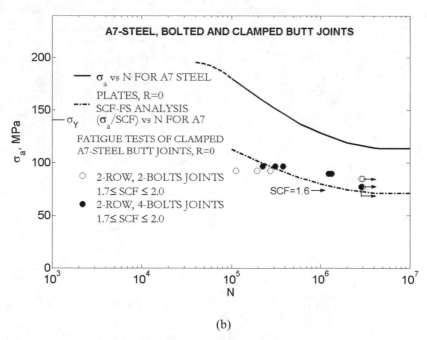

(b)

Figure 4.6. Comparison of fatigue test data for clamped, A7 steel butt joints with the predictions of the SCF-FS analysis [73–75].

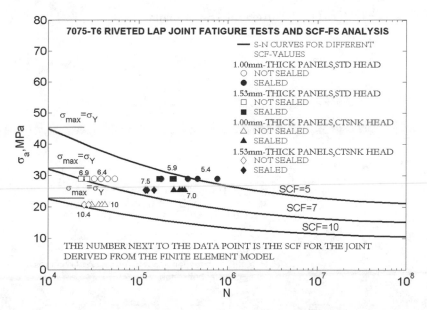

Figure 4.7. Influence of a low modulus, polymer sealant on the endurance of 7075-T6 aluminum, 3-rivets, 1 rivet-row lap joints (92 mm-wide, $P_1/D = 5$, t = 1.0 and 1.53 mm). Results of fatigue tests (R = 0.1) are compared with the SCF-FS analysis [6–8]. The test results are presented as discrete data points. The numbers next to the data points are the SCF-values derived from the following finite element models: 4S-8, 4SA-6, 4S-10, 4SA-7, 4C-9, 4C-25, 4C-11, 4CAC-27. The solid lines are predictions of the SCF-FS analysis for particular SCF-values.

TABLE 4.1. INFLUENCE OF INTERFERENCE AND HOLE EXPANSION ON THE FATIGUE STRENGTH OF ALUMINUM LUGS [78–80]

Lug Material & Dimensions[1]	%I[2]	Fatigue Strength[3]	Reference
7079-T6 Aluminum[4]	0.10	38	[77, 78]
a=18.3 mm	0.23	46	
d=15 mm	0.29	54	
t=9.0 mm	0.37[6]	--	
DTD363A[5]	0.2	28	[77, 79]
a=44.45 mm	0.4	34	
d=19.05 mm	0.56[6]		
t=9.02 mm	0.6	47	
	1.0	56	

[1] a-distance from center of hole to end of lug
  d-pin or bush external diameter
  t-thickness of lug
[2] Interference produced by steel pins or bushings. %X≈%I
[3] Fatigue Strength is stress amplitude for $N=10^6$ cycles
[4] 7079-T6 Aluminum, σ_Y=401 MPa, σ_u=446 MPa, 40 MPa$\leq\sigma_m\leq$132 MPa
[5] DTD363A Aluminum, $\sigma_Y\geq$481 MPa, σ_u=540 MPa and $\sigma_Y\geq$522 MPa, σ_u=587 MPa, σ_m=112 MPa
[6] Estimate of the value of %I corresponding to the onset of yielding

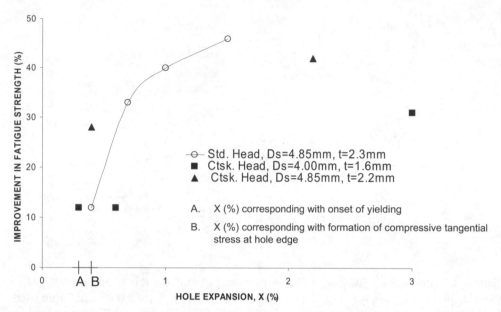

Figure 4.8. Improvements of the fatigue strength (stress amplitude at $N = 3 \times 10^5$ cycles) of 3 rivet-row, 2024-T3 lap joints produced by hole expansion after Muller [43] and Szolwinski and Farris [46,47].

of interference (or hole expansion), i.e. < 0.4, that preclude the development of compressive residual stresses and local yielding. In these cases, the improved fatigue strength can be linked with the reduction in the net section SCF as illustrated in Figure 3.8.

The variation of the increase of the fatigue strength of lap joints with hole expansion produced by upset and squeeze reported by Muller [34] and Szolwinski and Farris [46,47] are summarized in Figure 4.8. In this case, increases of the fatigue strength are not obtained until the hole expansions exceeds %X = 0.3% to 0.4% which correspond with the onset of yielding and the formation of compressive tangential stresses at the hole edge. It appears that the relaxation of residual stress that accompanies the release of the squeeze force—a relaxation absent when expansion is produced by interference—reduces the benefits of hole expansion. Figure 4.8 shows that the fatigue strength increases rapidly with hole expansions in the range 0.3<%X<0.7. The smaller improvements obtained with larger expansions may be related to the increasing amount of clamping attending hole expansions %X > 0.7.

The improvements produced by upset and squeeze hole expansion involve stress states that cannot be characterized with an elastic SCF (see Chapter 3.3). As a temporary expedient, an empirial strength reduction factor (SRF) is used to here to express the relation between the material S-N curve and that of the joint[4]. The SRF-values appropriate for multiple row lap joints fabricated with different amounts of hole expansion and tested by

---

[4] SRF = $FS_{MATERIAL}/FS_{JOINT}$

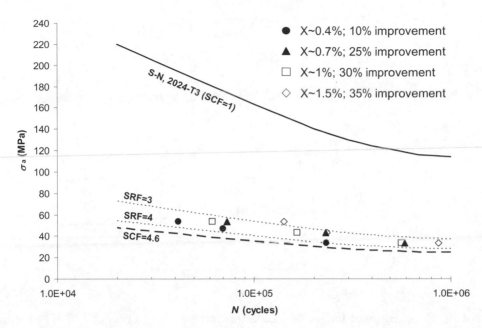

Figure 4.9. Fatigue strength of 2024-T3 aluminum lap joints fastened with rivets installed with different amounts of upset and squeeze leading to different amounts of hole expansion after Szolwinski and Farris [46,47]. The elevations of the fatigue strength produced by hole expansion are expressed in terms of the diminished values of the strength reduction factor (SRF).

Szolwinski and Farris [46,47] are identified in Figure 4.9. The SRF-values for a particular hole expansion vary with stress amplitude and increasing amounts of local yielding similar to the deviations from a fixed SCF noted in Figure B.1.

SRF-FS Analysis of Aluminum Lap Joint Panels Fastened with Self-Piercing, Steel Rivets. The finite element analyses of lap joints fastened with self-piercing rivets (Chapter 3.5) model the *shape* of the joint and the fasteners after piercing. They do not treat the residual stresses introduced by the piercing operation or subsequent plastic deformation during loading. The authors believe the stress-strain states generated in these joints are analogous to those produced by hole expansion and clamping (Chapter 3.3) and, unlike the F.E. models, do not lend themselves to characterization by an (elastic) SCF. For this reason, Figure 4.10 employs the strength reduction factor (SRF) introduced in the previous section to describe the joint performance and the SCF to characterize the behavior of the finite element models.

The fatigue tests of 5750-O Aluminum lap joints fastened with self-piercing steel rivets, taken together with the S-N curve for the material [44,81], reflect strength reductions in the range 5 < SRF < 7, as shown in Figure 4.10. The stress concentration factors, SCF = 12.2, 13.7 and 16.7, derived from the finite element analyses are more than two times these values and grossly understate the fatigue strength of the joints. The discrepancy emphasizes the inadequacy of the elastic SCF when yield-level residual stress-strain states are installed in the joint.

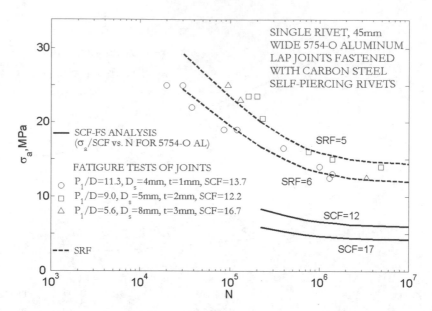

Figure 4.10. Comparison of fatigue test data for narrow, single fastener 5754-O aluminum lap joints fastened with self piercing, carbon steel rivets [81] with the predictions of the SCF-FS Analysis [44].

4.4 FATIGUE STRENGTH OF FASTENERS

Estimates of the stress amplitudes generated in riveted and bolted butt and lap joints are summarized in Tables 2.8 and 3.5, and relevant generalizations are offered in Chapters 2.8 and 3.1. The fatigue test results for aluminum and steel joints in Figures 4.3 to 4.6 represent cases where the failures proceeded in the joint rather than in the fasteners. Estimates of the cyclic stresses generated in the panels and fasteners of joints similar to those of Figures 4.3–4.6, and for comparable conditions, are listed in Table 4.2. The two examples offer insights into the factors contributing to the extended fatigue of the fasteners in these cases.

Aluminum Lap Joint. The riveted aluminum lap joint described in Figure 4.4 is similar to the ones for which test data is supplied in Figures 4.3 and 4.4. The joint is subjected to a relatively high mean stress. The estimated peak *panel* stress amplitude, $\sigma_a{}^* = 86$ MPa is close to the fatigue strength of the panel material corrected for the peak mean, FS ~ 80 MPa, The equality implies a panel life close to $N = 10^6$ cycles. The estimated peak *fastener* stress amplitude, $\sigma_a{}^* = 28$ MPa, is below the estimated *fastener* fatigue strength, FS~74 MPa (corrected for the $\sigma_m{}^* = 140$ MPa mean stress). This is consistent with a fastener life $N > 10^6$ and the finding that the fasteners outlast the panels.

Clamped Steel Butt Joint. The panels are subjected to a peak stress amplitude close to the fatigue strength of the steel panels in Figure 4.5 consistent with a panel life of $N = 10^6$ cycles. The clamping subjects the bolt to a yield stress-level mean stress, $\sigma_m{}^* = 845$ MPa. According to the Gerber diagram, for a yield-level mean stress (~90% of the ultimate

**TABLE 4.2. ESTIMATES OF THE PANEL AND
FASTENER STRESS AMPLITUDES AND FATIGUE
STRENGTHS FOR JOINTS AND CONDITIONS
COMPARABLE TO THOSE OF FIGURES 4.3 AND 4.5**

JOINT TYPE	Aluminum Lap[1]	Steel Butt[2]
MODE	Bearing	Clamping
%Cl	0	0.85
%ε	0	0.41
σ_s, MPa	0	849
PANEL STRESSES AND FATIGUE STRENGTHS		
σ_m, MPa	65	0
σ_m*, MPa	280	0
σ_a, MPa	20	100[3]
σ_a*, MPa	86	170
FS@N=10^6 cycles[4]	75 FS 95[5]	150[6]
FASTENER STRESSES AND FATIGUE STRENGTHS		
$\sigma_{a,33}$*$_{,TILT}$, MPa[7]	21	0
$\sigma_{a,33}$*$_{,TILT+SCF}$, MPa[8]	28	0
$\sigma_{a,33}$*$_{,BENDING}$, MPa	0	0[9] 60[10]
σ_m*, MPa[11]	208	849
FS@N=10^6 cycles[3]	~74	~60[12]

[1] Wide joint with 2 rows of rivets, P_1=30.6 mm, D_S=6.12 mm, H=3.06 mm
[2] Wide 1:1.6:1 joint with 2 rows of fasteners, P_1=66.9 mm, D_S=19.1 mm, H=28.6 mm
[3] Stress applied to center panel
[4] Fatigue strength adjusted for applied mean stress
[5] The 2 values quoted are for 2024-T3 and 7075-T6, respectively
[6] Approximate value for A514 and A440 steel panels
[7] Peak value produced in shank by fastener tilting, $\sigma_{a,33}$*=3$F_{a,33}$/ πD_S^2
[8] Estimate of peak value at head-shank junction and thread root stress
concentration, σ_a*=3$\sigma_{a,33}$*
[9] For bolted joint with shank-hole clearance
[10] For riveted joint with no clearance
[11] Peak mean stress
[12] For hardened steel nut and given level of mean stress

strength) the hardened bolt still retains about 20% of its fatigue strength, or FS ~60 MPa. This is more than adequate for a joint with clearance for which the peak stress amplitude is vanishingly small and accounts for the survival of the bolts. However, the peak stress amplitude generated by a tight fitting bolt or rivet is close to the FS and could this could lead to occasional fastener failures. In the absence of clamping and for a lower applied stress (σ_a = 50 MPa instead of 100 MPa, see Table 2.8), a hardened bolt is expected to survive, but the highly stressed panel will fail prematurely.

4.5. FRETTING WEAR AND FATIGUE

Fretting Wear. Repeated loading of structural shear joints produces fretting: micro-scale cyclic slips (relative to-and-fro movement) at the interfaces between panels and fasteners. Depending on the contact pressure, slip amplitude, surface finish and lubrication and the number of load cycles, fretting can produce significant amounts of abrasive wear. The

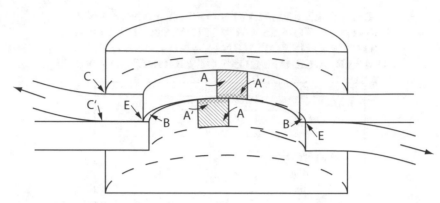

Figure 4.11. Locations in a shear joint subject to fretting damage. Locations A (for butt joints) and A' and B (for lap joints) at the shank-hole periphery are active in the bearing mode. Location C and C' in butt joints and C' in lap joints are active in the clamping mode.

generation of wear particles is promoted when slipping regions are exposed to high contact pressure along with high tangential tractions. The locations of the interfaces susceptible to fretting wear in joints with standard head fasteners have a bearing on NDE and are illustrated in Figure 4.11. The inclined portion of the head on the leading side of countersunk fasteners is also exposed to high pressures and large slips. The amount of wear expressed by the depth of the wear scar is proportional to the fretting wear parameter, F_1, which is defined in Appendix C. Figure 4.12 illustrates, for the bearing mode, the variations of F_1 in a lap joints around the periphery of fastener hole for different amounts of clamping and interference. The largest amounts of wear are to be expected in the locale of the peak F_1-values which are obtained in the vicinity of the net-section location. For joints in the clamping mode, the largest values of contact pressure are obtained directly under the fastener head at the head-panel interface, particularly at the leading edge of the head. Large pressures are also obtained at the panel-panel faying interfaces under the fastener head. Peak values of F_1 and the corresponding depths of the wear scar are summarized in Table 4.2. For stress amplitudes that provide relatively long fatigue lives, $N \sim 10^5$ cycles, the estimated depths of wear scars are in the range 0.4 μm < y < 3μm at locations A and A', and 3 μm < y < 30 μm at location C. Under these conditions fretting wear *in the absence of corrosion* is not expected to be a serious problem.

Fretting Fatigue. The process by which fretting reduces the fatigue life of contacting parts, called fretting fatigue, is described more fully Appendix C. The locations in shear joint interfaces where conditions may favor fretting fatigue are identified in Figure 4.10. Many workers, including the present authors and Birkemoe et al [53,54] have observed fretting wear debris at the bolt head-panel interface of steel butt joints (location C), and conclude from this that fretting "contributed" to the fatigue failure of these joints. The conclusion is open to question because fretting is not always accompanied by crack initiation.

To shed more light on the possible role of fretting fatigue, the values of the fretting fatigue parameter, F_2, at locations A, A' and C (Figure 4.11) have been calculated for butt and lap joints. The origin, use and limitations of F_2 as a rough, order-of magnitude indicator of the likelihood of fretting fatigue is discussed more fully in Appendix C. The variation of F_2 at

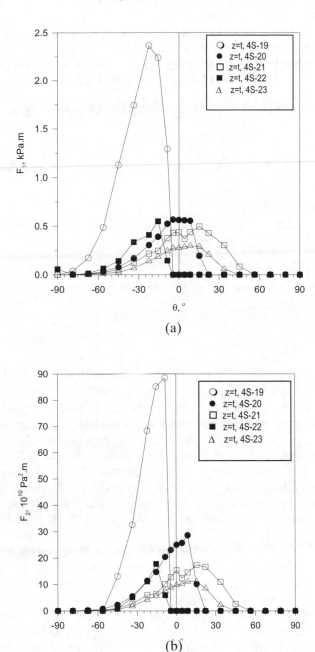

Figure 4.12. Variation of the fretting parameters with angular location and amounts of interference and clamping for wide, single-rivet-row, aluminum lap joints with standard head rivets (Models 4S-19, 4S-20, 4S-21, 4S-22 and 4S-23): (a) F_1 fretting wear parameter and (b) F_2 fretting fatigue parameter.

TABLE 4.3. ESTIMATES OF THE PEAK VALUES OF THE FRETTING WEAR, F_1, AND FRETTING FATIGUE PARAMETER, F_2, FOR JOINTS[1] AND CYCLIC STRESSES OFFERING EXTENDED CONVENTIONAL FATIGUE LIVES

CONDITIONS				LOCATION	F_1	Depth of Wear Scar, g, μm	F_2	$F_2 \div$Threshold F_2^2
Mode	σ_a, MPa	%I	%Cl	(Figure 4.8)	Pa-m	for N=10^5 cycles	10^{10} Pa^2-m	
Aluminum Butt Joint with Standard Rivet Heads								
Bearing	20	0	0	A	7	0.7	0.02	0.05
Bearing	20	0	0	C	12	1	0.03	0.07
Clamping	20	0	0.5	C	65	7	0.15	0.4
Aluminum Lap Joint with Standard Rivet Heads								
Bearing	20	0	0	A^1	16	2	0.10	0.2
Bearing	20	1	0	A^1	4	0.4	0.03	0.07
Bearing	20	0	0	C, C'	57	6	0.45	1
Clamping	20	0	0.5	C, C'	30	3	0.26	0.6
Aluminum Lap Joint with Countersunk Rivet Head								
Bearing	20	0	0	A^1	26	3	0.33	0.8
Bearing	20	1	0	A^1	4	0.4	0.03	0.07
Steel Butt Joint with Standard Bolt Head								
Clamping	50	0	0.85	C, C'	265	30	0.85	0.2

[1] P_1=30.6 mm, P_1/D_S=5
[2] F_2—fretting fatigue thresholds: Aluminum 4.2•10^9 Pa^{2}m, Steel 50•10^9 Pa^{2}m

the panel periphery with angular location is illustrated in Figure 4.12b. The peak values of F_2 generated by aluminum and steel butt and lap joints for 2-row joints and stress amplitudes that normally assure a long fatigue life are listed in Table 4.3. Values for butt and lap joints at locations A and A' at the periphery of the panel hole are below the "understated"[5] threshold for fretting fatigue. Hole expansion (interference) markedly reduces the severity of fretting at the panel hole periphery. Note that under these conditions, crack initiation does not usually occur at the hole periphery, but in the interior. The findings indicate that fretting fatigue at the A and A' locations is unlikely in butt and lap joints with standard head fasteners but more likely with countersunk rivets in the absence of hole expansion.

For the same local value of F_2, conditions for fretting fatigue of clamped panels with standard fastener heads are believed to be more favorable at C, where the rubbing of the fastener leading edge is accompanied by a steeper pressure and tangential stress gradient, than at C' at the faying surfaces. Similar conditions are obtained in lap joints with standard heads in the absence of clamping as a result of panel bending[6]. High contact pressures and tangential stresses are not obtained at the leading edge of a countersunk rivet which is more compliant. Table 4.2 shows that the F_2-values generated by clamped aluminum and steel butt joints are below the threshold. The F_2-values for the clamped and unclamped aluminum lap joint are close to or just exceed the threshold, respectively. For the special conditions evalu-

---

[5] The threshold values apply to the fretting conditions produced by a steep pressure gradient at a sharp edge similar to the conditions under the edge of the fastener head at location C (Figure 4.11). They "understate" the threshold at the periphery of the fastener-hole interface, locations A and A', where the attending pressure gradients are shallow by comparison.

[6] There is no gradient in butt joints in the bearing mode because the contact pressure between the head and panel top surface is negligible.

ated here and in the absence of hole expansion, the conditions at location C pose a marginal risk of fretting fatigue of lap joints in the bearing mode and both butt and lap joints in the clamping mode.. Three examples of fretting fatigue documented in the literature illustrate the difficulty of making generalizations about the conditions favoring fretting fatigue.

1. **Fretting Fatigue of a 3-row, countersunk riveted lap joint[7] [34].** The fasteners were installed with a very large squeeze force (hole expansion, $\%X \approx 2.6$) which produces a high, peak contact pressure at the faying surface, $p \approx 100$ MPa. Fretting conditions at the exposed surface of the "front" panel are not severe because the contact pressures under the leading edge of a countersunk head are not intense. Intense contact pressures are produced under the driven head of the same rivet. However the tangential stresses and slips generated at this location—the *lagging* row of holes of the "back" panel are reduced. Consistent with this, Muller [34] observed fretting fatigue at the *faying* surfaces on the leading side of the fastener at location C. The joint displayed a life of $N = 4.8 \times 10^5$ cycles at a stress amplitude of $\sigma_a = 46$ MPa and nearly two-fold increase in life when a 0.1 mm PTFE sheet is inserted between the faying surfaces. However, the PTFE sheet failed to improve the life of joints with smaller amounts of hole expansion and correspondingly smaller contact pressures, presumably, because the role of fretting fatigue was diminished in these cases.

2. **Fretting fatigue of a double strap, bolted butt joint[8].** Sharp, Nordmark and Menzemer [37], report that the unpainted butt joints joints supported a stress amplitude of $\sigma_a = 45$ MPa ($R = 0.33$) for $N = 10^7$ cycles. A zinc chromate primer applied to the *faying* surfaces increased the fatigue strength 60%. The relatively high stress amplitudes supported by this joint require large values of bolt tension and a high level of clamping The reason fretting proceeded at the faying surfaces is not clear. Relatively thick cover panels (their thickness was not disclosed) would have shifted the peak tangential stresses to the center panel.

3. **Fretting fatigue of a 3 rivet, single rivet-row lap joint[9].** In this case, Fretting fatigue proceeded at the faying surfaces (location B, Figure 4.11) where the sharp, curled edge of the fastener hole (at E) rubs against the adjacent, interior panel face [82]. The curled edge has the potential of producing a high contact pressure and the steep pressure gradient that promote fretting fatigue. Figure 4.13 shows that the accompanying fretting wear produces a (dark) crescent-shaped damage zone at the periphery of the fastener hole. Cracks initiated near the $-26°$ and $+206°$ locations where the fretting damage and the net section stress concentration overlap, locations relevant for NDE. This example of fretting fatigue involves special circumstances. The lap joint was subjected to a relatively high stress amplitude, $\sigma_a = 53$ MPa and peak stresses[10] that produced local yielding at the net section stress concentrations. Figures 4.11a and b illustrate that the yielding elongated the fastener holes

---

[7] $P_1 = P_2 = 20$ mm, $D_S = 4$ mm, and $t = 1$ mm.

[8] Unpainted, and zinc chromate primer painted 7075-T651 joint with 6.35 mm-thick test sections. Particulars about the type of bolts, bolt tension, joint dimensions or the relative thickness of the center panel and cover panels are not reported.

[9] 91.8 mm-wide, 3-countersunk rivets, single rivet-row, 7075-T6 lap joint, $P_1 = 30.6$ mm, $P_1/D = 5$, $t = 1.53$mm; $\sigma_a = 53$ MPa, $R = 0.1$ with test interrupted after $N = 3261$ cycles.

[10] $\sigma_{MAX} = 2$ SCF $\sigma_a = 690$ MPa.

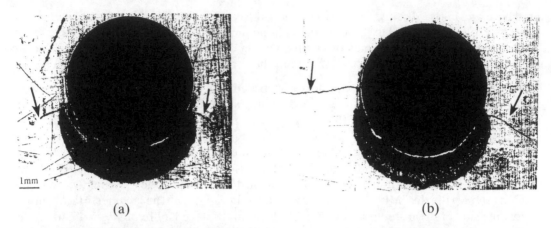

(a)   (b)

Figure 4.13. Examples of fretting damage and fatigue cracks produced in 7075-T6 aluminum, 3-rivet, single rivet-row lap joints (σ_a = 53 MPa, R = 0.1 and N = 3261 cycles), (82). The dark, crescent shaped regions below the fastener hole are produced by the fretting at the panel-panel interface (location B, Figure 4.11). The B-mode of fretting is responsible for the 3 cracks that emerge from the fretted region. The location of the 4th crack (left side of 19b) is consistent with the conventional fatigue at the net section location. The fact that the conventional fatigue crack, left side of (b), and the fretting fatigue crack, right side of (b), are nearly the same length is a sign that fretting fatigue did not curtail the life of the panel.

11%. This facilitated a 15-fold increase in the panel-panel slip which formed the wide, crescent-shaped damage zone that overlaps with the net section stress concentration. This mode of fretting fatigue did not reduce the fatigue life significantly (see Figure 4.12 caption).

The first two examples and the high F_2-values in Table 4.3 point to clamping, either from bolt tightening or rivet squeeze, as an important contributing factor. Overloading and the loosening of the joint are complicit in the third example. In each case fretting fatigue proceeded at the faying surfaces even though contact pressures and traction gradients are more intense at the panel surface directly under the edge of a protruding fastener head. In each case the joints were subjected to relatively high stress amplitudes. It should also be noted that the SCF-FS predictions presented in Figures 4.3–4.6 do not account for fretting fatigue. Yet, with possibly one exception, they do not overstate the fatigue strength of the joints. While there is little doubt that fretting fatigue reduces the fatigue strength of some joints, it is not a problem common to all joints.

DESIGN CONSIDERATIONS FOR REPEATED LOADING

5.1. MECHANICAL PERFORMANCE

Shear joints subject to cyclic loading are designed to satisfy three types of mechanical requirements:

1. **Monotonic Strength.** Joints must withstand the infrequent but extreme load conditions without gross yielding or fracture. Design procedures for assuring *ultimate strength* are described by Kulak [51], Bickford [40] and in other sources [37,83–85].
2. **Durability or Fatigue Life.** Joints must survive the number of stress cycles and the amplitudes and mean stresses they will be subjected to during their design life. In the past, the design specification for *durability and fatigue life* has been derived from fatigue tests of candidate joints, which is a fully empirical approach. The SCF-FS analysis provides a semi-empirical approach that can reduce the number of tests required for preliminary design. But the authors do not regard it as substitute for experimental verification. The application of the method, its limitations and validation are covered in Chapters 4.3 and 5.2 and Appendix B.
3. **Damage Tolerance.** *Damage Tolerance* is the capability of detecting growing fatigue cracks and repairing or replacing damaged parts before the performance of the joint is impaired. It involves matching the joint design with appropriate NDE procedures and practical inspection intervals. The joint design must provide a crack growth life of duration sufficient for a number of inspections. The number and their scheduling are chosen so that the sum of the probabilities of crack detention corresponding with the instantaneous crack lengths exceeds unity. To accomplish this, the variation of the crack length with number of stress cycles during the crack growth life must be defined. Analytical methods for forecasting crack growth life are treated by Fawaz [55] and others [52,56,57,59–64].

5.2 CONSIDERATIONS FOR JOINT DURABILITY

The selection of a joint configuration capable of supporting a particular stress amplitude and mean stress for a specified number of stress cycles can be facilitated with the SCF-FS

Analysis (Chapter 4.3). The analysis can be applied to both the panels and the fasteners, although SCF-values appropriate for fasteners are not as well defined in this text. It is appropriate for joint configurations, materials and stress conditions for which there is no prior service experience. It does require that the S-N curves of the panel and fastener material (for unnotched samples) be available. The method involves a number of uncertainties, identified later in this section. In view of this, joint configurations selected with the SCF-FS analysis should be regarded as preliminary and require experimental verification, particularly when the public health and safety are involved.

Panel Configuration. One of two approaches for selecting a viable panel configuration can be employed:

Approach 1. The starting point is the selection of the features of a particular type of shear joint that define the joint SCF including P_1/D, D/t, the number of rivet-rows, fastener type and load transfer mode. In addition, the service conditions are defined by the cyclic load and stress ratio, R, to be supported by the joint, and the desired cyclic life, N. The nominal stress amplitude the joint can support is defined by the material S-N curve (see Figure 5.1a): σ_a (N, joint) = σ_a (N, material)/SCF. Together with the load amplitude, the stress amplitude defines the gross cross sectional area, and combinations of panel width and thickness. One or more iterations may be required to obtain a panel thickness consistent with the initially assumed D/t-value. Finally, the dimensions are modified to provide an appropriate factor of safety.

Approach 2. In this case, the starting point is the selection of the stress amplitude the joint is intended to support. Again, the cyclic load and stress ratio, R, to be supported by the joint, and the desired cyclic life are identified. Together, the load amplitude and the stress amplitude define the gross cross sectional area, and tentative combinations of panel width and thickness. The maximum joint SCF that can be tolerated is defined by the material S-N curve (see Figure 5.1b): SCF = σ_a (N, material)/σ_a (N, joint). The next step is the selection of a particular type of shear joint including P_1/D, D/t, the number of rivet-rows, fastener type and load transfer mode, etc., that satisfies the SCF requirement. Width and thickness are then adjusted to accommodate the P_1/D, and D/t of the joint. Finally, the dimensions are modified to provide an appropriate factor of safety.

Fastener Configuration. The selection of a preliminary joint configuration also defines the fastener load amplitude, Q_a, and diameter to length ratio, the D_S/H. Following the procedure outlined in Tables 2.8 and 4.2, these parameters are converted into an estimate of the peak fastener stress amplitude, $\sigma_a{}^*$. The condition for fastener survival—the mean stress-modified fastener FS exceeds the $\sigma_a{}^*$—is then tested. When the fastener material/configuration does not pass the test, appropriate adjustments in joint configuration or load transfer mode are made and an acceptable joint and fastener are obtained by iteration.

Although a safety factor is always used to account for uncertainties, the following known sources of variation peculiar to structural shear joints and the fatigue process in general must be kept in mind.

1. **Assembly Variability.** The SCF-values quoted in this text are calculated for idealized assemblies. Variations in the SCF due to dimensional and assembly process variations such as hole diameter, fastener hole misalignment and buckled rivets on a multi-fastener joint are not known.

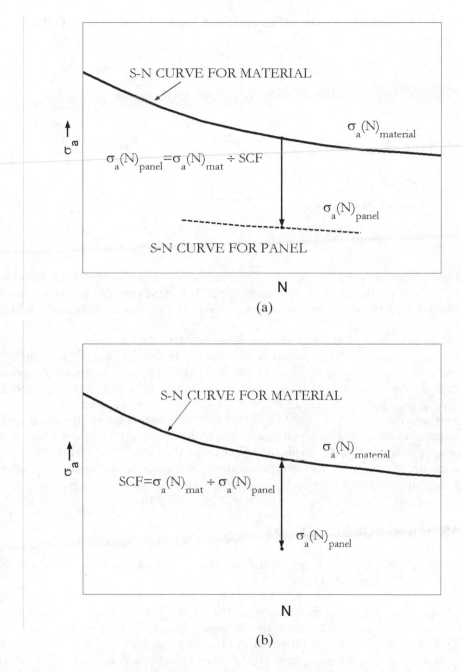

Figure 5.1. Schematic representation of the estimation procedures for selecting a joint panel configuration with a specified fatigue strength: (a) Approach 1 and (b) Approach 2.

TABLE 5.1. COMPARISON OF BUTT JOINT AND LAP JOINT FEATURES

PROPERTIES	BUTT JOINT		LAP JOINT	
	BEARING MODE	CLAMPING MODE	BEARING MODE	CLAMPING MODE
WEIGHT	2 X overlap + 2 X # fasteners		overlap + # fasteners	
CAPACITY OF CLAMPING		$\sim\sigma_c$		$2\sigma_c$-$3\sigma_c$
CYCLIC STRESS CAPACITY				
1. Panels with standard head fasteners	L	H^{st}	L	M
2. Panels with countersunk fasteners	L	H	L^{st}	L
3. Fasteners with clearance	$M^{24,130}$	L^{st0}	$H^{130,265}$	L^{st0}
3. Fasteners without clearance	$M^{24,130}$	$L^{3,29}$	$H^{130,265}$	$M^{24,39}$
BENEFIT OF MULTIPLE ROWS	H	N	H	M
BENEFIT OF ADHESIVE	H	L	H	M
BENEFIT OF HOLE EXPANSION	H	L	H	L
DETECTABILITY OF CRACKS (NDE)	L^{st}	L	L^{st}	L
FRETTING DAMAGE	L^{st}	L	M	M

L^{st} – lowest, L – low, M – moderate, H – high, H^{st} - highest

2. **Material Variability.** The S-N curves of structural materials display significant lot-to-lot variability. This is a problem because the available S-N curve measurements are rarely for the same lot as the panel material and because the minimum fatigue strength is rarely defined.

3. **R-Ratio.** The S-N curve of the material must correspond with the mean stress or stress ratio, R applied to the joint[1]. Material S-N data for the R-value of interest is frequently not available. In that case, an estimate of the material S-N curve for the appropriate R-value can be derived from a Gerber diagram (see Figure B.2). However, the approximate nature of the Gerber relation can introduce an uncertainty as large as 20%.

4. **The Cyclic Life.** The SCF-FS analysis is valid for relatively low values of stress amplitude when the peak stress levels within the stress concentrations do not exceed the yield strength: $\sigma_{MAX} = \sigma_{MIN} + 2\sigma_a < \sigma_Y$. This condition is obtained for $N > 10^5$ to $N > 10^6$ cycles, a range sometimes referred to as high-cycle fatigue. Estimates of the fatigue strength of joints for shorter cyclic lives (low-cycle fatigue) tend to be conservative.

5.3 JOINT TYPE AND MODE SELECTION

General features of butt joints and lap joints that affect their selection for service under fatigue conditions are summarized in Table 5.1.

Bearing Mode. For the bearing mode, comparable single-fastener row, 1:2:1 butt and lap joints (same panel thickness, P_1/D-ratio, fastener design, etc) harbor nearly the same SCF and are expected to display similar fatigue lives. The SCFs of butt and lap joints can be reduced by reducing pitch and adding additional fastener rows at the cost of more fasteners and more overlap, The major difference is that lap joints require half the overlap and half the fasteners of an equivalent butt joint and are therefore lighter. Another difference is that the risk of fretting fatigue is somewhat lower for butt joints than for equivalent lap joints.

---

[1] $\sigma_m = \sigma_a(1 + R)/(1 - R)$ and $R = \sigma_{MIN}/\sigma_{MAX}$.

This advantage is partially offset by the stiffer shank-hole interactions (higher k/K-values) in butt joints which elevate the 1st-row SCF of multiple row joints.

Clamping Mode. In the case of clamped joints, butt joints possess greater cyclic load carrying capacity than lap joints. In the fully clamped state, the SCF of the single fastener row butt joint is about half the value displayed by the equivalent lap joints. This translates into a two-fold cyclic load carrying advantage for the butt joint that can compensate for the extra overlap and fasteners. It should be noted that additional fastener rows do *not* improve the fully clamped butt joint, but extra fastener rows are expected to reduce the SCF of the clamped lap joint. Another feature unique to the clamped, 1:2:1 butt joint is that the residual gross section tensile stress (mean stress) of the cover panels is higher than the value for the center panel. The fatigue lives of the cover panels are correspondingly shorter. Optimum utilization of material is obtained from a joint with relatively thicker cover panels, e.g. from a 1.1:2:1.1 butt joint.

Adhesive-Plus-Bearing Mode. The results presented in Chapters 2.5 and 4.3 illustrate that the fasteners combined with adhesive offer benefits comparable to the clamping mode. While the findings in these chapters are confined to lap joints, similar improvements are anticipated for butt joints.

Fastener Cyclic Stress Capacity. The peak stresses generated in butt joint fasteners in the *bearing mode* are much smaller than those generated in the corresponding panel. The cyclic stresses produced in lap joint fasteners are much higher and comparable to the peak panel stresses. The clamping mode offers a four-fold reduction in the peak cyclic stresses applied to fasteners installed *without* clearance, and a nearly complete isolation from cyclic stresses for fasteners *with* clearance. However, clamping does generate high levels of mean stress in both butt and lap joint fasteners.

5.4 NDE AND FRETTING

Detection of Fatigue Cracks. The stress concentrations of 1:2:1 butt joints are located at the faying, interior surfaces of the front or back cover panel[2]. The detection of cracks formed in the back cover panel from the vantage point in front of the joint is particularly difficult. To "image" the crack, the NDI probe must penetrate the fastener head, front cover panel and center panel, as well as one fastener-panel and two panel-panel interfaces. The problem is almost as difficult for lap joints, where the stress concentrations are located at the interior surface of the front and back panel. In this case, the NDI probe must penetrate the fastener head, the front panel and the fastener-panel and two panel-panel interfaces to detect a crack in the back panel. The detection problem is less severe for lap joints with countersunk fasteners, which restrict the stress concentrations to front panel.

The fatigue crack detection problem for butt and lap joints can be eased at the cost of cyclic life by reducing the thickness of the front panels in the overlap region. In this way the

---

[2] The terms "front" and "back" are used to distinguish between the panels that can and cannot be visually inspected, respectively, when only one side of the joint is accessible (See Figure 1.2).

front panel SCF can be elevated above the levels displayed by the center or back panel. However, the design must assure that the growth lives of fatigue cracks that form in center and back panels always exceed the cyclic life of the joint.

Fretting. Fretting debris is invariably present near already obscured interfacial cracks. The generation of fretting debris increases as a crack grows due to higher local compliance and slip amplitude. The presence of fretting debris in the vicinity of a growing crack can significantly reduce the signal-to-noise ratio of the NDI measurement and therefore its reliability.

The success of the SCF-FS (which neglects wear) for elastic conditions, is a sign that fretting wear has a minor effect on the fatigue strength for large numbers of cycles; $N > 10^5$. For failure under elastic-plastic conditions, which corresponds to a relatively small number of slip reversals ($10^3 - 10^5$ cycles), the influence of fretting wear over and above that of cyclic macroplasticity is not known. Definitive understanding of the importance of fretting wear in joints with clamping and interference stresses remains to be developed.

5.5 FAIL SAFE DESIGN

Butt and lap joint panels are vulnerable to failure along the leading row of fasteners or just ahead of it. Such fractures are catastrophic for a shear joint because they completely sever the load bearing cross section. Adding additional fastener rows reduces the stresses at the leading row, but this row is still the most highly stressed and the joint remains vulnerable to catastrophic failure. Multiple row joints do offer the possibility of exchanging some joint life for a fail safe feature. This is obtained with a design that assures that the fasteners at the leading row fail before the panel. Failure of the fasteners does not compromise the panel; the load is simply transmitted to the remaining rows. The nearest row of fasteners assumes the role of the fasteners of the leading and lagging row in a joint with one less row. Although this row must support elevated rivet loads and SCFs, Table 2.5 shows that the demands on the panel at the promoted row are only a little higher than those at the original leading row. While this makes it likely that the promoted row will ultimately suffer the same fate, the fatigue life of the joint will be extended substantially allowing for timely failure detection and repair.

5.6 GENERALIZING FINITE ELEMENT RESULTS

The results of the finite element analyses presented here are mainly for single fastener row, aluminum butt and lap joints with one configuration: $P/D = 5$, $D/t = 4$. The analyses were performed for idealized rivet-like fasteners and examine a limited number of nominal stress levels. The results of analyses that employed linear elastic behavior can provide useful estimates of the SCFs, stress levels, strains and displacements for steel and other joint materials, bolted joints and a much wider range of shear joint configurations and nominal stresses.

Fastener Pitch, Diameter and Panel Thickness. Estimates of the SCF can be obtained for joint configurations in the range $4 < P_1/D_S < 8$, and $2 < D_S/t < 8$, using the relations between these ratios and the SCF derived by Seliger [32] and Fongsamootr [33] in Figures 2.9, 2.10 or 5.2. These relations were obtained for the bearing mode. The P_1/D_S dependence was

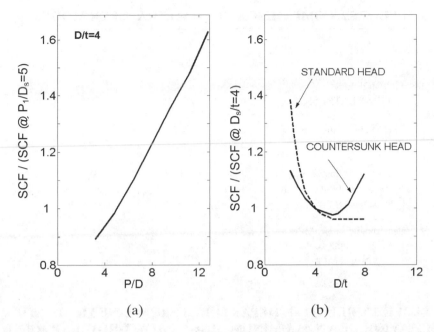

Figure 5.2. Correction factors for estimating the stress concentration factors, SCFs, of butt and lap joints for $3 < P_1/D_S < 13$ and $2 < D_S/t < 8$ given values for $P_1/D_S = 5$ and $D_S/t = 4$, (a) variation of correction with P_1/D_S after Seliger [32] and variation of correction with D_S/t after Fongsamootr [33].

obtained for single rivet-row lap joints, but is expected to approximate the behavior of multiple-row joints as well. The D_S/t-dependence applies only to lap joints and doublers. Relations similar to those in Figure 5.2 may apply to the clamping mode when the dependence is expressed in terms of the pitch-to-fastener head diameter ratio, P_1/D_B.

Number of Rivet-Rows. Analyses for single fastener-rows can be extended to joints with two, three, four or more rows, using expressions derived from the two springs model and the "superposition" approximation described in Chapter 17.

Nominal Stress. Table 5.2 illustrates that for joints with the same elastic material properties, the peak stress scales with the nominal stress to a close approximation and the SCF remains approximately constant. Contact pressures scale within about 10%, excess compliance, fastener tilt and various displacements scale less reliably. The departures arise from and depend on the relative contributions of interface slip and elastic strain to joint distortion. The positive departures indicate that the relative amounts contributed by interface slip diminish as the nominal stress is increased.

Applications to Steel and Other Panel Materials. Departures from linear elastic behavior of a steel lap joint relative to an otherwise identical aluminum joint are summarized in Table 5.3. The peak stresses and contact pressures for the steel and aluminum joints are almost

TABLE 5.2. COMPARISON OF LOAD TRANSFER MODES

			RELATIVE DEPARTURE FROM LINEARITY, %
F.E. MODEL	4S-8	4S-1	4S-8, 4S-1
NOMINAL STRESS, MPa	65	125	125
Excess compliance, m/GN	32.98	27.46	-17
Rivet tilt, degrees	2.51	3.6	-34
In-plane rivet-panel slip,μm	8.98	16.47	-4.8
In-plane panel-panel slip,μm	48.4	91.3	-1.9
Out-of-plane slip, μm	-3.7	-5.9	-21
Peak tensile stress, MPa	396	767	-0.7
Peak shank-hole contact			
pressure, MPa	453	779	-12

TABLE 5.3. RELATIVE DEPARTURES FROM LINEAR ELASTIC BEHAVIOR OF AN ALUMINUM, SINGLE RIVET-ROW LAP JOINT, $P_1/D = 5$, $D/t = 4$, ARISING FROM INTERFACE SLIP

			RELATIVE DEPARTURE FROM LINEARITY, %
F.E. MODEL	4S-1	4S-5	4S-1, 4S-5
JOINT MATERIAL	aluminum	steel	steel
NOMINAL STRESS, MPa	125	125	125
Excess compliance, m/GN	28.5	11.72	22
Rivet tilt, degrees	3.6	1.32	8.5
In-plane rivet-panel slip,μm	15.35	6.39	23
In-plane panel-panel slip,μm	94.3	32.1	0.8
Out-of-plane slip, μm	-6.4	-2.8	30
Peak tensile stress, MPa	761	766	0.7
Peak shank-hole contact			
pressure, MPa	859	866	0.8

identical and display no departure. This means that estimates of these quantities derived from calculations for the identical aluminum joint, corrected for differences in nominal stress, are reliable. The excess compliance, tilt and certain displacements of the aluminum joint, modified by the aluminum-to-steel elastic modulus ratio, understate the values for the steel joint by as much as 30%. These discrepancies arise because the relative contributions of interface slips to displacement values are larger in the otherwise stiffer steel joint. More accurate estimates can be made by taking the departures into account.

Rivets vs Bolts. The proportions of the rivet with "standard" head are similar to those of standard hexagonal bolts:

	RIVET	HEXAGONAL BOLT
Shank diameter-to-head height ratio,	1.6	1.5
Fastener head diameter-to-shank diameter ratio,	1.6	1.5

Results obtained for joints fastened with rivets with the "standard" head approximate joints fastened with snug fitting bolts.

SECTION 2

COMPILATION OF FINITE ELEMENT RESULTS

LOAD TRANSFER IN SINGLE RIVET-ROW LAP JOINTS (CONVENTIONAL AND COUNTERSUNK)

6.1 SUMMARY OF LOAD TRANSFER MODELS

The finite element models, 4S-1, 4C-1 and 4S-5 (see Tables A.2 and A.3), used to obtain these data are described in detail in Chapter 13; the material properties for the various models are described in Chapter 15. The content of this chapter is taken from References [2] and [3].

6.2 LOAD TRANSFER CALCULATIONS FOR SINGLE RIVET-ROW LAP JOINTS

Lap joints are different from butt joints in that significant out-of-plane distortion occurs in the form of rivet tilting and panel bending. As a result there are out-of-plane load transfer paths not found in an idealized two-dimensional butt joint analysis. These can be seen in Figs. 6.1(a)–6.1(c). Figure 6.1(a) shows a loaded, single rivet-row lap joint with conventional rivets; the distortion is exaggerated for clarity. Figures 6.1(b) and 6.1(c) show free body diagrams of the upper halves of the conventional and countersunk rivets and the corresponding panel pairs with the forces that act on them. The four load transfer paths will be the loading direction components of the following forces:

(1) the shear across the rivet midsection, (N2 + F1), shown in the top part of Fig. 6.1(b),
(2) the axial force on the rivet shank midsection, (N1 + F2), shown in the top part of Fig. 6.1(b),
(3) the friction force between the upper and lower panel, (F3), shown in Fig. 6.1(c), and
(4) the normal force between the upper and lower panels, (N3), shown in Fig. 6.1(c).

In addition, the rivet and the rivet head must support large axial forces:

(1) the axial rivet shank load at the head (equal to the shear force at the rivet head), (N1), shown in Fig. 6.1(c), and
(2) the axial rivet shank load at the midsection, (N1+F2), shown in the top part of Fig. 6.1(b).

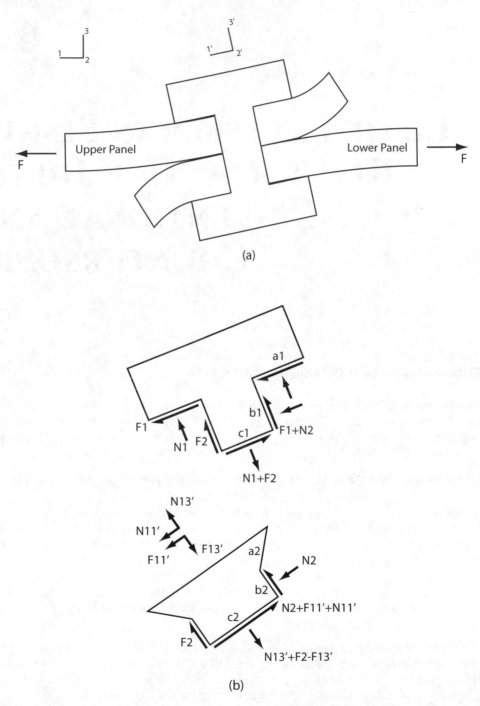

Figure 6.1. (a) A loaded single-rivet lap joint (distortions are exaggerated for clarity), (b) Free body diagram of the upper half of a standard head (top) and a countersunk head (bottom) rivet. See 6.1(a) for prime definitions. N = normal force, F = friction force, (c) Free body diagrams of the two sets of panels *(figure continues)*.

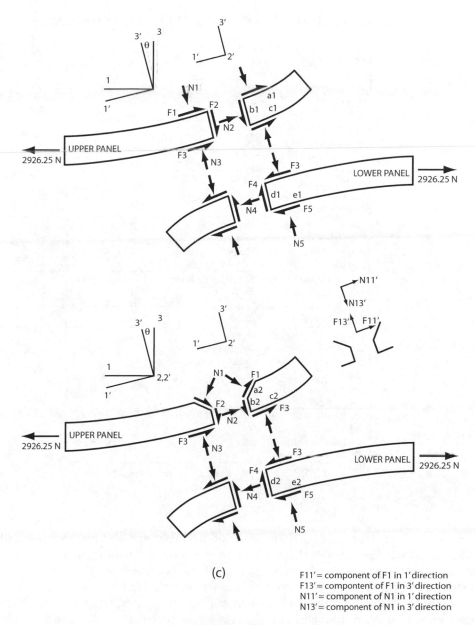

(c)

F11′ = component of F1 in 1′ direction
F13′ = compontent of F1 in 3′ direction
N11′ = component of N1 in 1′ direction
N13′ = component of N1 in 3′ direction

Figure 6.1. *Continued.*

The local and global axes are defined in Fig. 6.1(a), and Figures 6.1(c) and 6.2 identify the various contact surfaces. The local (primed) axes represent the directions normal and tangent to the tilted rivet. At each contact area (*a1, b1, c1, d1, e1* for Models 4S-1 and 4S-5, and *a2, b2, c2, d2, e2* for Model 4C-1) there is a normal force and friction force acting at the panel surfaces. The local normal and friction forces represent the contact pressure and the contact

TABLE 6.1. DEFINITIONS OF EVALUATED FORCES

Variable	Definition	Surface	Direction
N1	Local out-of-plane normal force	*a1*	3'
F1	Local in-plane friction force	*a1*	1'
N2	Local in-plane normal force	*b1*	1'
F2	Local out-of-plane friction force	*b1*	3'
N3	Local out-of-plane normal force	*c1*	3'
F3	Local in-plane friction force	*c1*	1'
N4	Local in-plane normal force	*d1*	1'
F4	Local out-of-plane friction force	*d1*	3'
N5	Local out-of-plane normal force	*e1*	3'
F5	Local in-plane friction force	*e1*	1'
U3'	Total local out-of-plane forces for 4S-1 (=N1+F2-N3)	upper panel	--
L3'	Total local out-of-plane forces for 4S-1 (=N5+F4-N3)	lower panel	--
U1'	Total local in-plane forces for 4S-1 (=F1+N2-F3)	upper panel	--
L1'	Total local in-plane forces for 4S-1 (=F3+N4-F5)	lower panel	--
U3	Total global out-of-plane force (L3 on lower panel) (=U1'sinθ+U3'cosθ=0)	upper panel	3
U1	Total global in-plane in-plane force (L1 on lower panel) (=U1'cosθ+U3'sinθ=2926.25 N)	upper panel	1
N11'	Component of N1, local in-plane normal force	*a2*	1'
N13'	Component of N1, local out-of-plane normal force	*a2*	3'
F11'	Component of F1, local in-plane friction force	*a2*	1'
F13'	Component of F1, local out-of-plane friction force	*a2*	3'
U3'	Total local out-of-plane forces for 4C-1 (=N13'+F2-N3-F13')	upper panel	--
L3'	Total local out-of-plane forces for 4C-1 (=N5+F4-N3)	lower panel	--
U1'	Total local in-plane forces for 4C-1 (=F11'+N11'+N2+F3)	upper panel	--
L1'	Total local in-plane forces for 4C-1 (=F3+N4+F5)	lower panel	--

shear acting on each contact area. These forces are evaluated and resolved into either the 1'-direction (local-in-plane) or the 3'-direction (local-out-of-plane). Then the local components are transformed into the 1-direction (global-in-plane) and the 3-direction (global-out-of-plane) using the angle of rivet tilt. The normal and friction forces at all contact surfaces are obtained by multiplying the average contact pressure or the average contact shear by the contact area. Free body diagrams of the forces acting on each panel are shown in Figs. 6.1. The forces acting at the panel ends are calculated from the tensile stress (σ) and cross-sectional area of the panels (F = 125 MPa * 1.53mm * 15.3 mm = 2926.25 N). The variables shown in Figs. 6.1 are defined in Table 6.1. At the countersunk portion of Model 4C-1, both the friction and normal forces have components in the local-in-plane and the local-out-of plane directions; therefore these forces must be resolved into the 1'- and 3'- directions.

Figures 6.2(a) and (b) show typical contour plots illustrating values of contact pressure distributions for the aluminum models 4S-1 and 4C-1.

Table 6.2 presents the local in- and out-of-plane loads obtained for the models. The load labels refer to the notation described in Figs. 6.1.

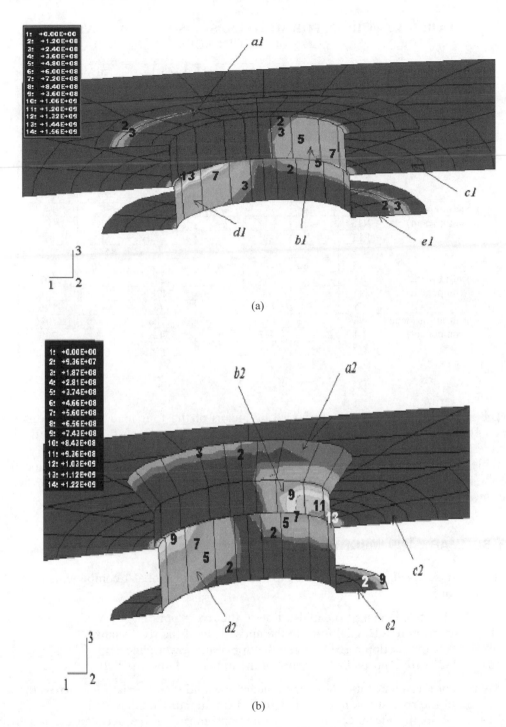

(a)

(b)

Figure 6.2. (a) Contour plot of contact pressure for Model 4S-1 at contact surfaces *a1*, *b1*, *c1*, *d1* and *e1*, (b) Contour plot of contact pressure for Model 4C-1 at contact surfaces *a2*, *b2*, *c2*, *d2* and *e2*, σ_{nom} = MPa.

TABLE 6.2. FORCES FOR MODELS 4S-5, 4S-1 AND 4C-1 AT ALL CONTACT SURFACES

Load (N)		Steel Model 4S-5 (standard rivet) rivet tilt=1.32°		Al Model 4S-1 (standard rivet) rivet tilt=3.6°		Al Model 4C-1 (countersunk) rivet tilt=4.24°	
load type	load label	upper panel	lower panel	upper panel	lower panel	upper panel	lower panel
Local-in-plane (3'-component)	F1	-152	-	-148	-	-	-
	F11'	-	-	-	-	-41	-
	N11'	-	-	-	-	-57	-
	N2	-2498	-	-2482	-	-2567	-
	F3	-268	-	-221	-	-207	-
Local-out-of-plane (1'-component)	N1	-781	-	-746	-	-	-
	N13'	-	-	-	-	-880	-
	F13'	-	-	-	-	114	-
	F2	-426	-	-542	-	-546	-
	N3	1378	-	1106	-	1036	-
Local-in-plane (3'-component)	F3	-	268	-	221	-	207
	N4	-	2442	-	2478	-	2499
	F5	-	154	-	148	-	196
Local-out-of-plane (1'-component)	N3	-	-1378	-	-1106	-	-1036
	F4	-	457	-	542	-	329
	N5	-	777	-	746	-	984

The applied tensile load is transferred by means of the four load transfer paths listed above. Table 6.3 shows the magnitudes and percentages of load transferred within the joint, for the models, through the four different load paths.

Table 6.4 shows axial rivet shank loads for the two aluminum models at three locations: (1) the head of the upper rivet, (2) the middle cross section of the rivet, and (3) the head of the lower rivet.

6.3. SUMMARY AND IMPORTANT POINTS

1. The load is transferred through the lap joint by the four global-1 components of the following forces:

   (a) 90–91% from shear at the midsection of the rivet shank
   (b) 1–3% from tensile axial force at the midsection of the rivet shank
   (c) 7–9% from friction force between the upper and lower panel
   (d) (–)1–3% from normal force between the upper and lower panel.

2. There is not a significant difference in shear force and tensile axial force across the midsection of the rivet shank in the standard and countersunk models.
3. There are small differences in friction force and normal force between the upper and lower panels for the standard and countersunk models because the standard rivet heads clamp the panels more effectively than the countersunk rivet head does. The steel model displays slightly higher friction and normal forces than the aluminum model.

TABLE 6.3. LOAD TRANSFER AT THE UPPER RIVET MIDSECTION AND UPPER PANEL OF ALUMINUM MODELS 4S-5, 4S-1 AND 4C-1

Load Transfer Path	Total Force (N)					
	local-plane(1'-3')*			global- in-plane(1-component)*		
	4S-5	4S-1	4C-1	4S-5	4S-1	4C-1
1.Shear force across rivet midsection (% of total applied force)	2650 (91%)	2629 (90%)	2665 (91%)	2650 (91%)	2624 (90%)	2658 (91%)
2.Axial force on rivet shank midsection (% of total applied force)	1207 (41%)	1288 (44%)	1312 (45%)	28 (1%)	81 (3%)	97 (3%)
3.Friction force between upper & lower panel (% of total applied force)	268 (9%)	221 (8%)	207 (7%)	268 (9%)	220 (8%)	206 (7%)
4.Normal force between the upper and lower panel (% of total applied force)	1378 (47%)	1106 (38%)	1036 (35%)	-32 (-1%)	-69 (-2%)	-77 (-3%)
5.Total force Sum of 1 through 4 (% of total applied force)	-	-	-	2914 (100%)	2856 (98%)	2884 (99%)

* see Figs. 6.1 for directions

TABLE 6.4. AXIAL RIVET SHANK LOAD FOR MODELS 4S-1 AND 4C-1 AT THE UPPER AND LOWER RIVET MIDSECTION

Rivet shank	Total force (N)	
	Model 4S-1	Model 4C-1
Upper rivet head	746	765
Upper rivet midsection	1288	1312
Lower rivet head	746	984
Lower rivet midsection	1288	1314

4. There is a higher force acting at the rivet shank upper portion, and at the rivet head lower portion, for the countersunk head model than for the standard head model. The peak stress occurs at these locations.
5. In spite of the fact that load transfer is essentially the same in the standard and counter-sunk head models, there is a large difference in the panel stress concentration factor (SCF = 6.1 and SCF = 9.8 for the standard head and countersunk head model, respectively).

CHAPTER

7

COMPILATION OF RESULTS FOR OPEN HOLE PANELS AND BUTT JOINTS

7.1 SUMMARY OF CALCULATIONAL MODELS FOR OPEN HOLE PANELS

Table A.5 presents the calculations done for open hole panels. The finite element models used to obtain these data are described in detail in Chapter 13; the material properties for the various models are described in Chapter 15. The content of this section is taken from Reference (1).

7.2 SUMMARY OF CALCULATIONAL MODELS FOR BUTT JOINTS

Tables A.1, A.7 and A.8 in Appendix A present the various calculations done for butt joints. The finite element models used to obtain these data are described in detail in Chapter 13; the material properties for the various models are described in Chapter 15. The content of this section is taken from References [1] and [9].

7.3 SUMMARY OF CALCULATION RESULTS FOR OPEN HOLE PANELS

Stress concentration factors (SCF), defined as the ratio of the maximum tensile stress to the nominal stress based on one repeat section, were evaluated for four different holes, one straight, three countersunk, and one straight hole with a pin. SCFs are listed in Table 7.1.

Figures 7.1 and 7.2 compare the stress distributions for an open hole panel with a panel containing a pin.

7.4 SUMMARY OF CALCULATION RESULTS FOR BUTT JOINTS WITHOUT INTERFERENCE

Table 7.2 presents a summary of the distortion parameters from a 2-D analysis of an elastic butt joint with headless fastener under a nominal, remote stress of 125 MPa, without interference and with $\mu = 0.2$.

TABLE 7.1. STRESS CONCENTRATION FACTORS DUE TO HOLE COUNTERSINKING

Model	Hole Geometry	SCF
1H-1	Straight hole	3.0
1H-2	100° countersink upto 1/2 the panel thickness	3.38
1H-3	61.6° countersink through the entire panel thickness	3.23
1H-4	100° countersink through the entire panel thickness	3.5
1H-5	Straight hole with pin	2.85

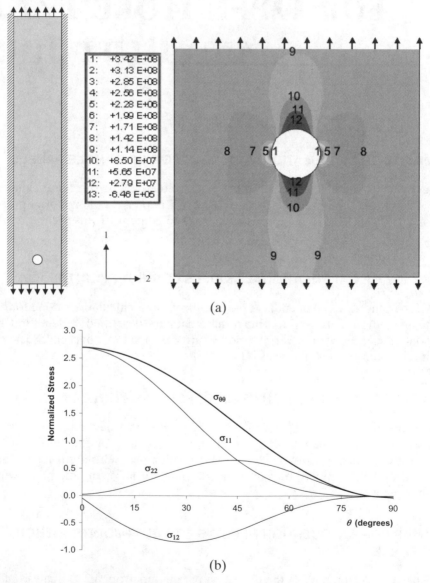

(a)

(b)

Figure 7.1. Long plate with open empty hole: (a) σ_{11} contours with plate bottom fixed, pin has modulus of sheet (aluminum) (b) Distribution of bulk stresses adjacent to the hole. Stresses are normalized by applied stress $\sigma_{nom} = 125$ MPa.

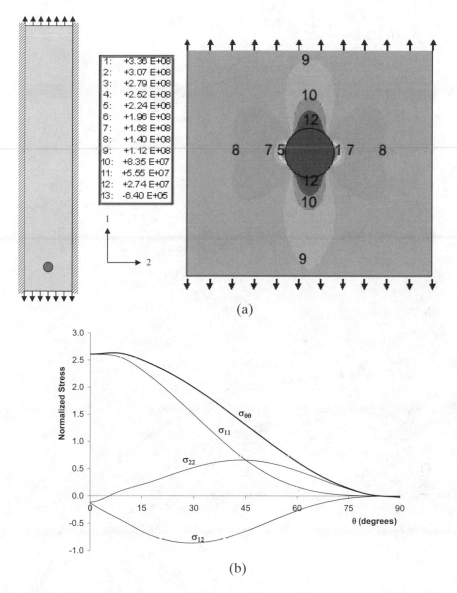

(a)

(b)

Figure 7.2. Long plate with open pin-filled hole: (a) σ_{11} contours with plate bottom fixed, pin has modulus of sheet (aluminum) (b) Distribution of bulk stresses adjacent to the hole. Stresses are normalized by applied stress $\sigma_{nom} = 125$ MPa.

The effects of the friction coefficient, μ, pin-plate material combinations, finite plate dimensions and biaxial loading on the normalized contact pressure, $p \cdot R/Q$, contact angle, θ_0, and the normalized tangential stress around the hole, $\sigma_{\theta\theta} \cdot R/Q$, are summarized in Figures 7.3 and 7.4. The results are precise for the conditions for which these calculations were performed (R = 3.06 mm, and Q = 1836 kN/m), and will be close for other values, but not precise, due to slight nonlinearities in the problem (as discussed in Chapters 1.2 and 5.6).

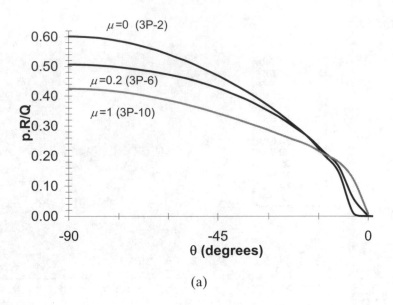

(a)

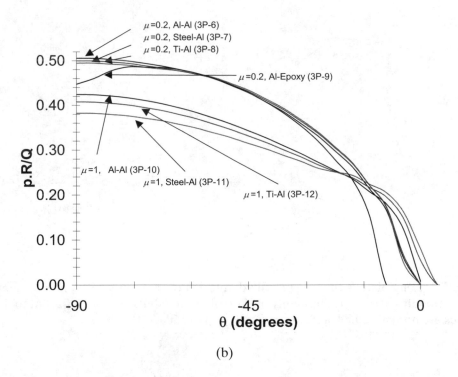

(b)

Figure 7.3. Normalized pin-plate contact pressure, p.R/Q, as a function of angular location, θ, along the hole circumference (see Figure 2.2(a) for angular location convention). (a) μ = 0, 0.2 and 1, infinite plate, aluminum alloy pin and aluminum alloy plate, (b) μ = 0.2 and 1, infinite plate, and four pin-plate material combinations *(figure continues)*.

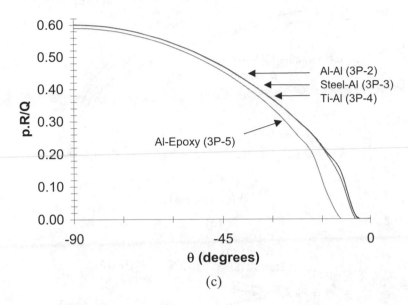

(c)

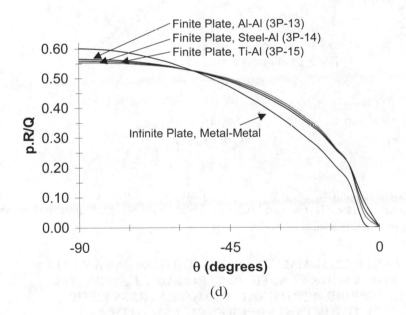

(d)

Figure 7.3. *(figure continued)* (c) $\mu = 0$, infinite plate, and four pin-plate material combinations, (d) $\mu = 0$, finite and infinite plate, and the metallic pin-plate material combinations considered in the study. Under these conditions, the contact pressure distribution is essentially a single curve *(figure continues)*.

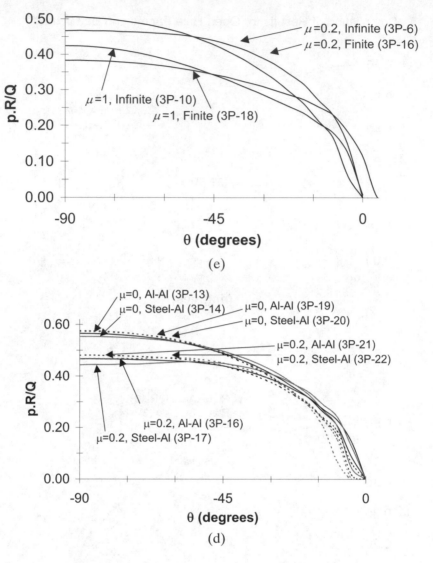

Figure 7.3. *(figure continued)* (e) μ = 0.2 and 1, finite and infinite plate, aluminum alloy pin and aluminum alloy plate, (f) μ = 0 and 0.2, finite plate, two pin-plate material combinations and two types of loading (uniaxial and biaxial).

**TABLE 7.2. SUMMARY OF DISTORTION PARAMETERS
FOR A WIDE ELASTIC PINNED BUTT JOINT, $P_1/D = 5$,
UNDER A NOMINAL, REMOTE STRESS 125MPa,
WITHOUT INTERFERENCE AND WITH μ = 0.2,
Q = 3825 kN/m, MODEL 3P-1, AFTER IYER [1]**

Feature	Value
Loading end displacement, μm	307.1
In-plane slip, μm	0-38
Peak contact pressure, MPa	522
Stress concentration factor (SCF)	6.4

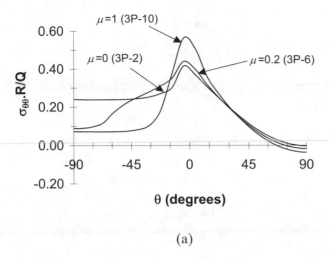

(a)

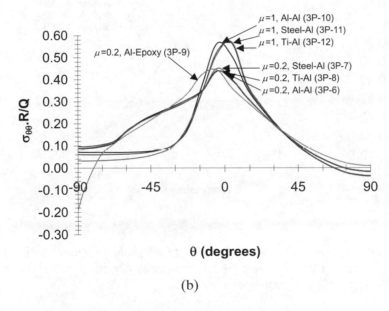

(b)

Figure 7.4. (a) Normalized tangential stress, $\sigma_{\theta\theta}.R/Q$, around the hole as a function of angular location, θ, along the hole circumference (see Figure 2.2(a) for angular location convention). (a) $\mu = 0, 0.2$ and 1, infinite plate, aluminum alloy pin and aluminum alloy plate, (b) $\mu = 0.2$ and 1, infinite plate, and four pin-plate material combinations. *(figure continues)*

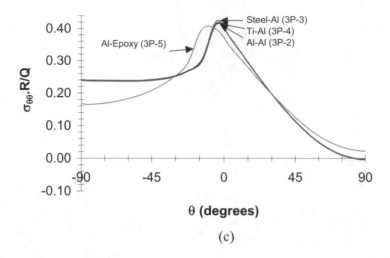

(c)

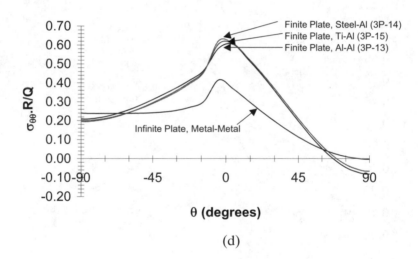

(d)

Figure 7.4. *(figure continued)* (c) $\mu = 0$, infinite plate, and four pin-plate material combinations, (d) $\mu = 0$, finite and infinite plate, and the metallic pin-plate material combinations considered in the study. *(figure continues)*

7.5 EFFECTS OF INTERFERENCE IN BUTT JOINTS

Figures 7.5(a) and (b) and 7.6 illustrate the angular variation of contact pressure about a pinned hole undergoing cyclic stress for three different values of friction and interference.

Figure 7.7 shows the slip amplitude, δ (the difference between the load and unload value of the local slip displacements), for an aluminum pin with 1% interference.

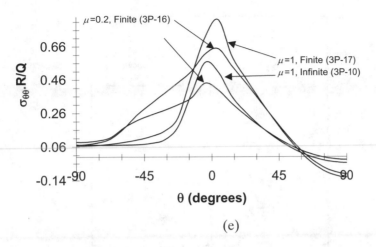

(e)

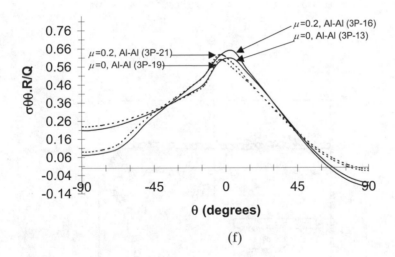

(f)

Figure 7.4. (e) $\mu = 0.2$ and 1, finite and infinite plate, aluminum alloy pin and aluminum alloy plate, (f) $\mu = 0$ and 0.2, finite plate, aluminum alloy pin and aluminum alloy plate, and two types of loading (uniaxial and biaxial).

Figures 7.8 and 7.9 present the angular variations of fretting wear parameters F1 and F2 for different values of interference and friction.

The peak values of F1 and F2, their angular positions and corresponding values of contact pressure and slip, are described in Figs. 7.10(a) and 7.10(b) and in Tables 7.3 and 7.4.

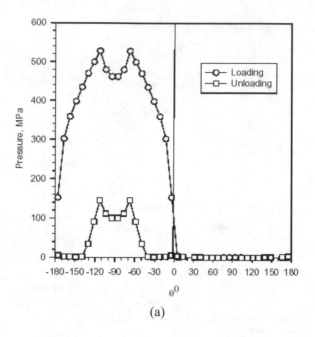

(a)

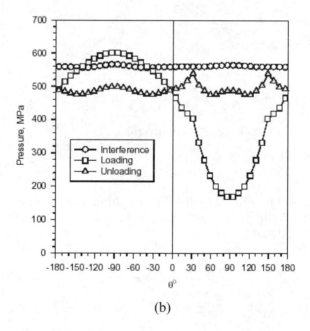

(b)

Figure 7.5. Angular variation of the contact pressure about the pinned hole with the cyclic stress applied and partially released for the aluminum pin: (a) $\mu = 0.5$ and 0% interference (b) $\mu = 0.2$ and 2% interference.

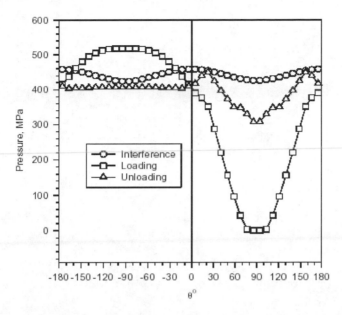

Figure 7.6. Angular variation of the contact pressure about the pinned hole with the cyclic stress applied and partially released for the aluminum pin, with $\mu = 0.5$ and 1% interference.

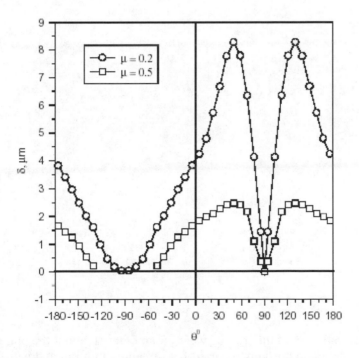

Figure 7.7. Angular variation of the slip amplitude about the pinned hole with the cyclic stress applied and partially released for the aluminum pin and 1% interference.

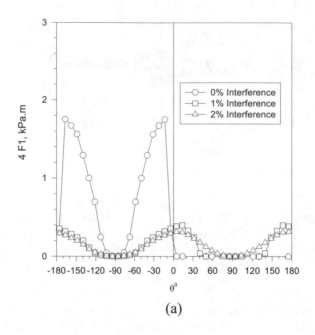

(a)

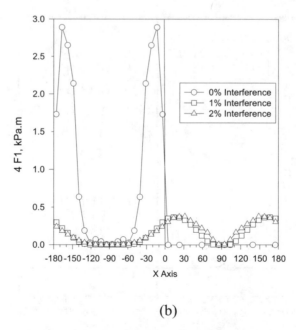

(b)

Figure 7.8. Angular variation of the fretting wear parameter, F1, about the pinned hole with the cyclic stress applied and partially released for the aluminum pin and with 1% and 2% interference: (a) $\mu = 0.2$ (b) $\mu = 0.5$.

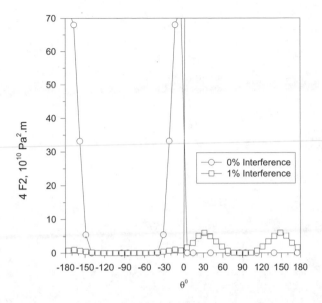

Figure 7.9. Angular variation of the fretting wear parameter, F2, about the pinned hole with the cyclic stress applied and partially released for the aluminum pin and with 1% and 2% interference, and $\mu = 0.5$.

Figure 7.11(a) illustrates a typical, calculated variation with angular location, θ, of the fretting wear and fatigue parameters F1, F2 and the tensile stress in the panel, adjacent and parallel to the hole. Figure 7.11(b) shows the variation of the hysteresis loop width around the pin-hole interface as a function of friction, load and interference.

Table 7.5 gives angular locations of peak values of F1 and F2 and circumferential tensile stresses adjacent to the pin-hole interface.

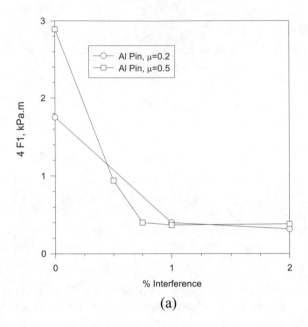

(a)

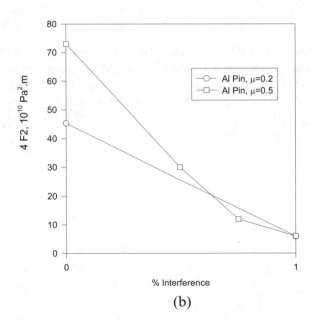

(b)

Figure 7.10. Variation of the fretting parameters with interference and friction coefficient: (a) F1 (b) F2.

TABLE 7.3. INFLUENCE OF INTERFERENCE ON THE ANGULAR LOCATION, θ, OF THE PEAK VALUES OF F_1, AND CORRESPONDING VALUES OF δ AND p AT THE PIN-HOLE INTERFACE FOR EIP BEHAVIOR

Pin Material, % Interference	Model	μ	F_1 (kPa.m)	θ range (^{0})	δ (μm)	p (MPa)
Al, 0	3P-25	0.2	0.44	-13.1	24.8	336.1
Al, 1	3P-26	0.2	0.10	13.7	4.8	416.7
Al, 2	3P-27	0.2	0.08	4.6	2.9	536.3
Al, 0	3P-28	0.5	0.72	-13.3	19.7	301.5
Al, 0.5	3P-29	0.5	0.24	13.7	7.9	244.0
Al, 0.75	3P-30	0.5	0.10	13.6	2.5	326.3
Al, 1	3P-31	0.5	0.09	13.6	2.0	376.3
Al, 2	3P-32	0.5	0.10	31.5	1.9	403.6

TABLE 7.4. INFLUENCE OF INTERFERENCE ON THE ANGULAR LOCATION, θ, OF THE PEAK VALUES OF F_2, AND CORRESPONDING VALUES OF δ AND p AT THE PIN-HOLE INTERFACE FOR EIP BEHAVIOR

Pin Material, % Interference	Model	μ	F_2 (10^{10} Pa^2.m)	θ range (^{0})	δ (μm)	p (MPa)
Al, 0	3P-25	0.2	11.33	-13.0	24.8	336.1
Al, 1	3P-26	0.2	1.48	13.8	4.8	416.7
Al, 2	3P-27	0.2	-	-	-	-
Al, 0	3P-28	0.5	18.25	-4.0	25.0	151.7
Al, 1	3P-31	0.5	1.50	31.5	2.3	287.1
Al, 2	3P-32	0.5	-	-	-	-

- The stresses in the panel, adjacent and parallel to the hole remain compressive and F_2 does not have any significance.

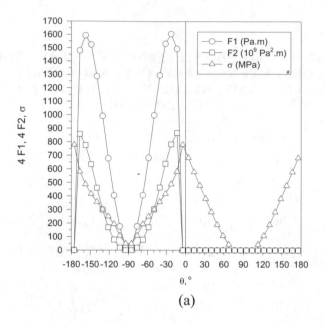

(a)

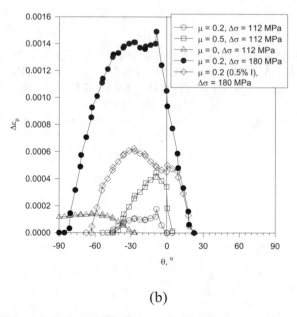

(b)

Figure 7.11. (a) Angular variation of F1, F2 and the tensile stress, σ, in the panel, adjacent and parallel to the hole interface for the case of μ = 0.2 and 0% interference (b) Angular variation of the hysteresis loop width around the pin-hole interface as a function of μ, load and interference.

TABLE 7.5. INFLUENCE OF INTERFERENCE ON THE ANGULAR LOCATION, θ, OF THE PEAK VALUES OF F_1, F_2 AND CIRCUMFERENTIAL TENSILE STRESSES ADJACENT TO THE PIN-HOLE INTERFACE FOR ELKP BEHAVIOR

Pin Material, % Interference	Model	μ	Δσ (MPa)	F_1 θ (°)	F_2 θ (°)	$\sigma_{\theta\theta}$ θ (°)
Al, 0	3P-33	0.0	112	-	-	-4.5
Al, 0	3P-34	0.2	112	-22.5	-13.5	-4.5
Al, 0	3P-35	0.3	112	-22.5↔-13.5	-13.5	-4.5
Al, 0	3P-36	0.5	112	-13.5	-13.5	-4.5
Al, 0.5	3P-37	0.2	112	-4.5	-4.5↔4.5	22.5
Al, 0	3P-38	0.2	180	-22.5	-13.5	-4.5
Al, 0.5	3P-39	0.2	180	-13.5	-4.5	4.5

COMPILATION OF RESULTS FOR LAP JOINTS

8.1 SUMMARY OF CALCULATIONAL MODELS

Single Rivet-Row Lap Joints. Tables A.2 and A.3 present a summary of the calculations done for single rivet-row lap joints with standard and countersunk heads, respectively. Table A.7 in Appendix A shows the calculations performed for evaluation of the effects of interference and clamping on single rivet-row lap joints. The finite element models used to obtain these data are described in detail in Chapter 13; the material properties for the various models are described in Chapter 15.

Double Rivet-Row Lap Joints. Table A.4 in Appendix A presents a summary of the calculations done for double rivet-row lap joints. The finite element models used to obtain these data are described in detail in Chapter 13; the material properties for the various models are described in Chapter 15.

The content of this chapter is taken from References [1,13–17,19,20,35].

8.2 SUMMARY OF CALCULATION RESULTS

Single Rivet-Row Lap Joints. Table 12.1 contains the results of nine calculations for single rivet-row lap joints. The data includes joint compliance, rivet tilt, total in-plane slip at the rivet-panel interface and the panel-panel interface, total out-of-plane slip, peak tensile stresses and peak contact pressure.

Table 8.1 lists calculated stress concentration factors and their angular locations for the five elastic models. The angular locations of the peaks are the likely sites of fatigue crack initiation when the fatigue process is not assisted by fretting. A normalized peak compression, NPC, defined as the negative ratio of the peak compressive stress to the nominal tensile stress based on one repeat section, is also presented.

**TABLE 8.1. CALCULATED STRESS
CONCENTRATION FACTORS IN THE PANELS
OF THE RIVETED ASSEMBLIES**

Model	Tensile SCF		NPC		
	value	θ range , °	value	θ, °	depth, mm
4S-1	6.1	-14.9° to 8.6°	-6.2	-90°	-1.53
4S-5	6.13	-14.9° to 8.6°	-6.39	-90°	-1.53
4C-1	9.8	-11.4° to 5.4°	-8.4	-90°	-0.765 to -1.53
4C-2	8.8	-13.5° to 11.2°	-7.2	-90°	-0.8 to -1.53
4C-5	10.5	-17.4° to 5.4°	-10.8	-90°	-1.53
4C-6	12.1	-12.7° to 0°	-15.6	-90°	-1.53

**TABLE 8.2. IMPORTANT STRESSES GENERATED
IN THE RIVET SHANK**

Model	Peak Shank-Hole Compressive Stress (σ_{11}), MPa	Peak Shear Stress (σ_{13}), MPa	Peak Axial Tensile Stress (σ_{33}), MPa	Peak Axial Compressive Stress (σ_{33}), MPa
4S-1	-438.9	239	309	-129
4S-2	-309.8	228	329	-116
4C-1	-628.7	325	418	-360
4C-2	-577.8	388	489	-257
4C-5	-608.5	289	617	-123
4C-6	-668.2	367	902	-94

Table 8.2 summarizes some of the important stresses generated in the rivet shank.

Figure 8.1 shows a zoomed-in profile of the displaced mesh and corresponding tensile stress contours for the single rivet-row lap joint.

Double Rivet-Row Lap Joints. Table 12.2 in Chapter 12 contains the results of calculations for double rivet-row lap joints. The data includes joint compliance, rivet tilt, total in-plane slip at the rivet-panel interface and the panel-panel interface, total out-of-plane slip, peak tensile stresses and peak contact pressure. The data presented are valid for elastic, double rivet-row, aluminum alloy lap joints with standard and countersunk aluminum and steel rivets. They are compared with similar analyses for single rivet-row joints.

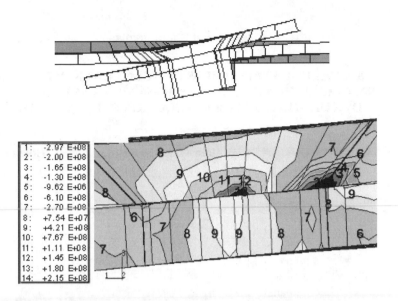

Figure 8.1. Zoomed in profile of the displaced mesh and corresponding tensile stress (σ_{11}) fields generated in the two holes in model 4C-6. In the latter, the rivet is not shown and displacements are magnified 3x in the displaced mesh profiles, σ_{nom} = 60 MPa.

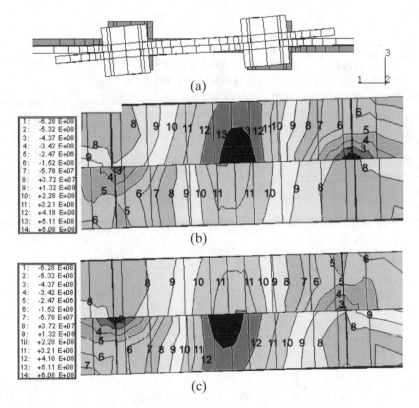

Figure 8.2. Zoomed in profiles of the displaced mesh of model 5S-1 and corresponding tensile stress (σ_{11}) fields (b) in hole #1 (c) in hole # 2. The shaded in region in (a) is the initial configuration and displacements are magnified 3x, σ_{nom} = 125 MPa.

**TABLE 8.3. CALCULATED STRESS CONCENTRATION FACTORS (SCF)
AND NORMALIZED PEAK COMPRESSION (NPC) VALUES IN THE PANELS
OF THE THREE-DIMENSIONAL RIVETED ASSEMBLY**

Model		SCF				NPC			
		Hole #1		Hole #2		Hole #1		Hole #2	
		Value	$\theta,°$	Value	$\theta,°$	Value	$\theta,°$	Value	$\theta,°$
Non-countersunk aluminum rivet 5S-1	upper panel	4.4	6.2 to -10.9	3.1	-3.1 to -8.7	-3.8	-90	-3.1	-90
	lower panel	3.1	8.7 to 3.1	4.4	10.9 to -6.2	-3.1	90	-3.8	90
Countersunk aluminum rivet 5C-1	upper panel	6.1	9.8 to -15.6	4.5	0 to -9.8	-5.8	-90	-4.0	-90
	lower panel	3.0	16.1 to -8.6	4.2	12.1 to -8.6	-2.7	90	-4.3	90
Countersunk steel rivet 5C-2	upper panel	5.9	10.7 to -13.2	4.2	8.2 to -15	-4.5	-90	-3.8	-90
	lower panel	2.8	8.7 to 2.5	4.0	8.6 to -3.1	-2.6	90	-4.1	90

**TABLE 8.4. IMPORTANT STRESSES GENERATED
IN THE RIVET SHANK**

Model	Peak Shank-Hole Compressive Stress (σ_{11}), MPa		Peak Shear Stress (σ_{13}), MPa		Peak Axial Tensile Stress (σ_{33}), MPa		Peak Axial Compressive Stress (σ_{33}), MPa	
	rivet #1	rivet #2	rivet #1	rivet #2	rivet #1	rivet #2	rivet #1	rivet #2
Non-countersunk aluminum rivet 5S-1	-240	-240	126	126	174	174	-95	-95
Countersunk aluminum rivet 5C-1	-353	-321	175	143	160	224	-115	-164
Countersunk steel rivet 5C-2	-338	-335	168	133	200	266	-116	-157

A displaced mesh for model 5S-1 is illustrated in Fig. 8.2.

Table 8.3 shows calculated values of the stress concentration factor (SCF), defined as the ratio of the maximum tensile stress to the nominal stress based on one repeat section, and its angular location at the two sets of holes in the model.

Figure 8.2 illustrates contours of important stresses developed in the rivets under load: (i) σ_{11}, arising from contact between the rivet shank and upper and lower panels, along the

TABLE 8.5. MAXIMUM SCF AND SAF VALUES
IN THE PANELS OF DOUBLE RIVET-ROW LAP JOINTS

Model	Hole #	Panel	Single Rivet-Row SCF = SAF	Double Rivet-Row SCF	Double Rivet-Row SAF
Non-countersunk aluminum rivet 5S-1	1	Upper	6.1	4.4	8.8
		Lower	6.1	3.1	6.2
	2	Upper	-	3.1	6.2
		Lower	-	4.4	8.8
Countersunk aluminum rivet 5C-1	1	Upper	9.8	6.1	12.2
		Lower	5.9	3.0	6.0
	2	Upper	-	4.5	9.0
		Lower	-	4.2	8.4
Countersunk steel rivet 5C-2	1	Upper	8.8	5.9	11.8
		Lower	5.4	2.8	5.6
	2	Upper	-	4.2	8.4
		Lower	-	4.0	8.0

loading axis, (ii) σ_{13}, shear stress present in the rivet, and (iii) σ_{33}, axial stress present in the rivet.

Table 8.4 lists peak values of the stresses generated in the rivet shank.

The SCF's obtained for the double-row assemblies are compared in Table 8.5 with those obtained for comparable single-row assemblies subjected to the same, remote, repeat distance load and nominal stress. Defining the "nominal" rivet load as the remote, repeat distance load divided by the number of rivet rows, a "nominal" Stress Amplification Factor (SAF), defined as the peak tensile stress divided by the product of the nominal rivet load and the remote repeat distance panel cross-sectional area, may be evaluated. The SAF (equal to SCF * N, where N is the number of rivet rows) and SCF are identical for a single rivet-row assembly. Table 8.5 contains values for the SAFs for the non-countersunk and countersunk cases.

The SCF's express the relation between the peak tensile stresses generated in the panel and the remote nominal stress. The values obtained for the double-row assemblies are compared in Table 8.6 with those obtained for comparable single-row assemblies subjected to the same, remote, repeat distance load and nominal stress.

To examine the origin of the variation in the distribution of in- and out-of-plane load transmission with the number of rivet rows, the peak rivet shear stress, σ_{13}, indicative of in-plane transmission, and the peak rivet axial stress, σ_{33}, indicative of out-of-plane transmission, are examined. These are listed in Table 8.7. Two parameters which are easily deduced and indicate the relative proportions of in- and out-of-plane load transmission are listed.

8.3 SUMMARY OF CALCULATION RESULTS INCLUDING INTERFERENCE AND CLAMPING

Single Rivet-Row Lap Joints. Table 8.8 shows the magnitudes of stresses and plastic strains generated in the upper panel, parallel and immediately adjacent to the hole surface as a function of depth, due to the installation step, i.e., interference or/and clamping.

TABLE 8.6. PANEL HOLE SCFS AND SAFS IN DOUBLE RIVET-ROW LAP JOINTS

Joint Type (Model)	Hole #	Panel	Single Rivet-Row	Double Rivet-Row	
			SCF = SAF	SCF	SAF
Non-Countersunk		Upper	6.1	4.4	8.8
Aluminum	1	Lower	6.1	3.1	6.2
Rivets		Upper	-	3.1	6.2
(5S-1)	2	Lower	-	4.4	8.8
Countersunk		Upper	9.8	6.1	12.2
Aluminum	1	Lower	5.9	3.0	6.0
Rivets		Upper	-	4.5	9.0
(5C-1)	2	Lower	-	4.2	8.4
Countersunk		Upper	8.8	5.9	11.8
Steel	1	Lower	5.4	2.8	5.6
Rivets		Upper	-	4.2	8.4
(5C-2)	2	Lower	-	4.0	8.0

Table 8.9 shows the magnitudes and gradients of stresses and plastic strains generated in the lower, non-countersunk hole due to the accommodation of 1% and 2% interference (models 4C-25 and 4C-26), and Table 8.10 shows the same values generated in the upper, countersunk hole because of the installation step.

Figures 8.3–8.5 show the variations with the angular location, θ, of the contact pressure, and the in-plane slip in the upper panel of the S-type models. Table 8.11 shows values of several

TABLE 8.7. COMPARISONS OF PROPORTIONS OF SHEAR AND ROTATION IN RIVETS IN DOUBLE ROW LAP JOINT MODELS

Model	Rivet #	Single Rivet-Row (MPa)		Double Rivet-Row (MPa)		σ_{13}/σ_{33}		$\dfrac{(\sigma_{13})}{((\sigma_{13})^2+(\sigma_{33})^2)^{0.5}}$	
		σ_{13}	σ_{33}	σ_{13}	σ_{33}	Single Row	Double Row	Single Row	Double Row
5S-1	1	239	309	126	174	0.77	0.72	0.61	0.59
	2	-	-	126	174	-	0.72	-	0.59
5C-1	1	325	418	175	160	0.78	1.09	0.61	0.74
	2	-	-	143	224	-	0.64	-	0.54
5C-2	1	388	489	168	200	0.79	0.84	0.62	0.64
	2	-	-	133	266	-	0.5	-	0.45

TABLE 8.8. FEATURES OF UPPER PANEL HOLE DEFORMATION DUE TO THE INSTALLATION STEP (INTERFERENCE AND/OR CLAMPING) IN THE NON-COUNTERSUNK ASSEMBLIES

Model	Depth	$\sigma_{\theta\theta}$, MPa	$p\varepsilon_{mag}$ (%)	$p\varepsilon_{\theta\theta}$ (%)
4S-20	z=0	35.0	0.244	0.201
	z=0.5t	113.5	0.218	0.212
	z=t	324.6	0.146	0.132
4S-22	z=0	-53.1	0.00	0.00
	z=0.5t	4.0	0.00	0.00
	z=t	-14.2	0.00	0.00
4S-23	z=0	-304.5	0.839	0.584
	z=0.5t	141.8	0.209	0.202
	z=t	279.8	0.208	0.170

TABLE 8.9. FEATURES OF THE NON-COUNTERSUNK (LOWER) PANEL HOLE DEFORMATION DUE TO THE INSTALLATION STEP (INTERFERENCE AND/OR CLAMPING) IN THE COUNTERSUNK ASSEMBLIES

Model	Depth	$\sigma_{\theta\theta}$, MPa	$p\varepsilon_{mag}$ (%)	$p\varepsilon_{\theta\theta}$ (%)
4C-25	z=t	310.2	0.10	0.10
	z=1.5t	130.2	0.19	0.18
	z=2t	73.0	0.15	0.14

TABLE 8.10. FEATURES OF THE COUNTERSUNK (UPPER) PANEL HOLE DEFORMATION DUE TO THE INSTALLATION STEP (INTERFERENCE AND/OR CLAMPING) IN THE COUNTERSUNK ASSEMBLIES

Model	Depth	$\sigma_{\theta\theta}$, MPa	$p\varepsilon_{mag}$ (%)	$p\varepsilon_{\theta\theta}$ (%)
4C-25	z=0	158.8	0.00	0.00
	z=0.5t	101.6	0.05	0.05
	z=t	269.3	0.09	0.09
4C-27	z=0	1.1	0.00	0.00
	z=0.5t	-0.5	0.00	0.00
	z=t	29.4	0.00	0.00

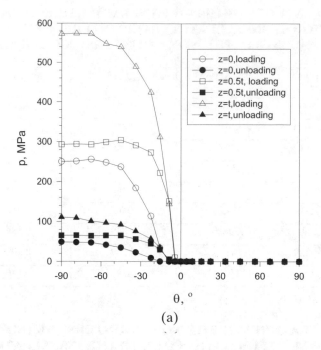

(a)

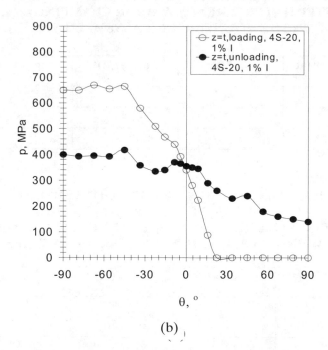

(b)

Figure 8.3. (a) Variation of contact pressure at the shank-hole interface with angular location and depth for Model 4S-19, (b) Variation of contact pressure with angular location at depth $z = t$, where contact pressures peak, for Model 4S-20.

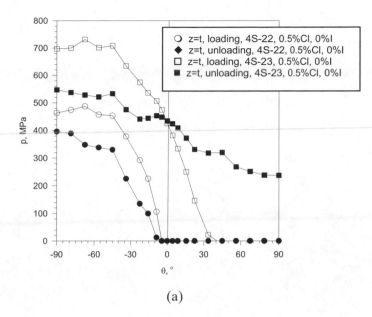

(a)

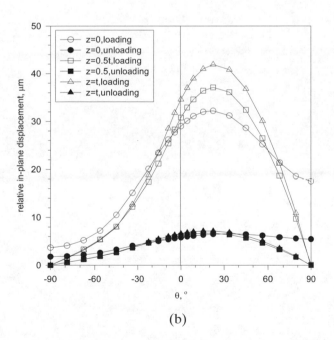

(b)

Figure 8.4. (a) Angular variation of contact pressure at the shank-hole interface in Models 4S-22 and 4S-23 at $z = t$, depth at which the pressures peak, (b) Variation of in-plane slip between the panel hole and rivet in Model 4S-19.

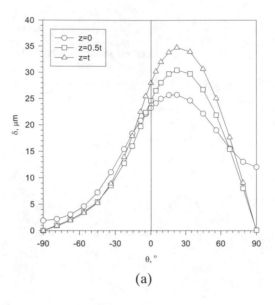

(a)

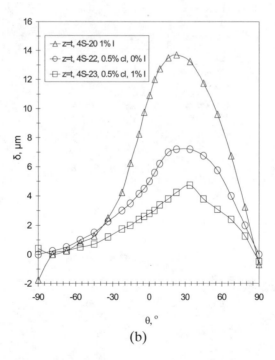

(b)

Figure 8.5. (a) Variation of in-plane slip amplitude at the shank-hole interface with angular location and depth for Model 4S-19, (b) Angular variation in Models 4S-20, 4S-22 and 4S-23 at depth $z = t$, where slip amplitudes peak.

**TABLE 8.11. VARIATION OF PARAMETERS
THROUGH THE DEPTH OF THE HOLE**

Model		Installation $\sigma_{\theta\theta}$ (MPa)	Peak $\sigma_{\theta\theta}$ (MPa)	Peak δ_{12} (μm)
4S-19	z=0	0	429.9	25.7
	z=0.5t	0	533.2	30.4
	z=t	0	594.8	34.7
4S-20	z=0	35	95.0	2.7
	z=0.5t	114	196.4	4.4
	z=t	325	518.3	7.2
4S-22	z=0	-53	337.8	9.3
	z=0.5t	4	372.1	12.2
	z=t	-14	410.2	13.7
4S-23	z=0	-305	-216.6	1.5
	z=0.5t	142	218.5	3.1
	z=t	280	411.1	4.8

*Installation $\sigma_{\theta\theta}$ values are from the Isotropic response analyses

parameters at different locations through the hole depth. Composite effects are expressed by the mechanistic parameters, F_1 and F_2, whose variations with angular location at z = t are shown in Fig. 8.6. Table 8.12 shows peak values of the mechanistic parameters through the depth of the shank-hole interface and their angular location.

For the upper panel of C-type models, Figs. 8.7–8.9 show the variations with angular location, θ, of the contact pressure and the in-plane slip. Table 8.13 shows values of several parameters at different locations through the hole depth. Variations of the mechanistic parameters, F_1 and F_2, with angular location at z = t are shown in Fig. 8.10. Table 8.14 shows the peaks of the mechanistic parameters through the depth of the shank-hole interface.

Double Rivet-Row Lap Joints. Table 8.15 shows the peak values of some parameters across the hole depth. Note that +90° corresponds to the loading end for upper hole #1 but −90° corresponds to the loading end for lower hole #2.

Figures 8.11–8.12 show variations of the contact pressure and in-plane slip, with angular location, θ, in the critical holes in the two models. Table 8.16 lists the peak values of the tangential shear stress, F_1 and F_2 (the primary fretting parameters) across the hole depth in the critical holes of the two models.

Figure 8.13 shows the angular variation of the local cyclic stress range, $\Delta\sigma$ and local mean stress, σ_m, at z = t in the critical holes. Variations with angular location at z = t for F_1 and F_2 are shown for the critical holes in Fig. 8.14.

Table 8.17 summarizes the critical fretting fatigue conditions in single- and double-row assemblies with and without interference.

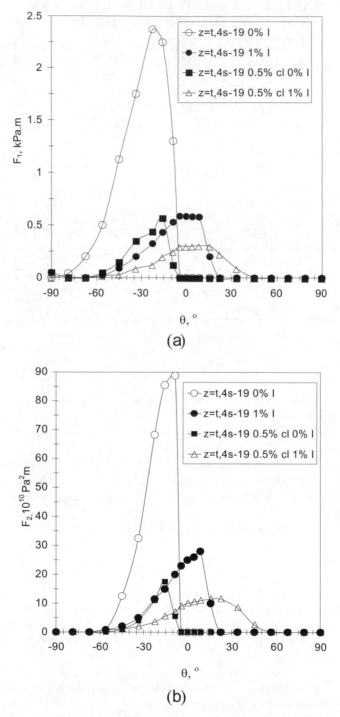

Figure 8.6. Variation of fretting parameters with angular location at $z = t$, the depth at which the parameters peak, for Models 4S-19, 4S-20, 4S-22 and 4S-23. (a) fretting wear parameter, F_1, and (b) fretting fatigue parameter, F_2.

**TABLE 8.12. SUMMARY OF MECHANISTIC PARAMETERS
INDICATING LOCATIONS OF FRETTING WEAR
AND CRACK INITIATION IN LAP JOINT ASSEMBLIES
WITH NON-COUNTERSUNK RIVETS**

Model		Peak Tangential Stress (MPa)		Peak F_1 (kPa.m)		Peak F_2 (10^{10} Pa^2.m)	
		magnitude (MPa)	θ, °	magnitude (kPa.m)	θ, °	magnitude (10^{10} Pa^2.m)	θ, °
4S-19	z=0	102	-45.0	0.20	-33.7	5.65	-22.5
	z=0.5t	116	-33.7, -45.0	0.35	-15.6, -22.5	15.08	-8.7
	z=t	137	-33.7, -45.0	0.57	-15.6, -22.5	22.13	-8.7
4S-20	z=0	83.1	15.6	0.05	15.6	0.35	0.0, 4.3
	z=0.5t	77	15.6, -4.3	0.07	33.7, 8.7	1.35	22.5, 15.6
	z=t	124	-4.3, -8.7	0.14	8.7, -8.7	7.25	8.7
4S-22	z=0	82.6	-45.0	0.08	-4.3, -15.6	2.38	-8.7
	z=0.5t	57.8	-33.7	0.07	-15.6	2.08	-15.6
	z=t	132	-33.7	0.14	-15.6	4.45	-15.6
4S-23	z=0	89	33.7, 22.5	0.03	33.7, 15.6	-	-
	z=0.5t	50	22.5, -4.3	0.03	33.7, 8.7	0.73	22.5, 15.6
	z=t	102	-4.3, -8.7	0.07	15.6, 0.0	2.80	15.6, 8.7

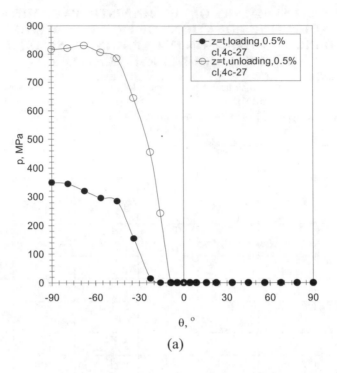

(a)

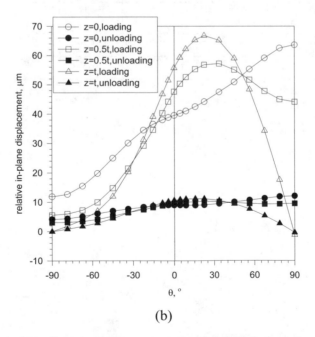

(b)

Figure 8.7. (a) Variation of contact pressure at the shank-hole interface with angular location and depth for Model 4C-24, (b) Angular variation of contact pressure at shank-hole interface at depth $z = t$, where contact pressures peak, for Model 4C-25.

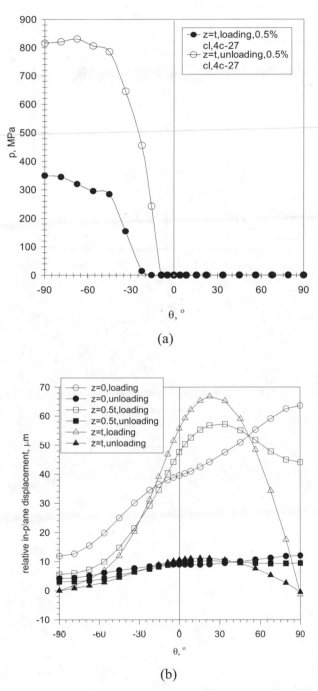

Figure 8.8. (a) Angular variation of contact pressure at the shank-hole interface at depth $z = t$, where contact pressures peak Model 4C-27, (b) Angular variation of in-plane slip between the panel hole and rivet in Model 4C-24.

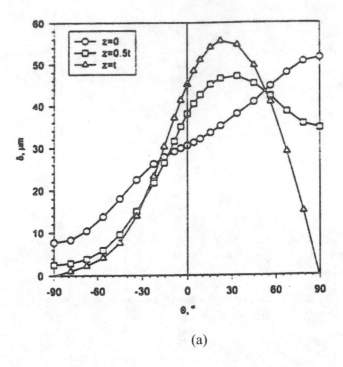

(a)

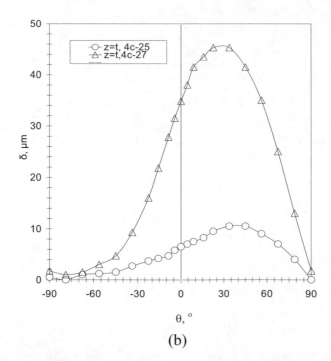

(b)

Figure 8.9. (a) Angular variation of in-plane slip amplitude in shank-hole interface in Model 4C-24, (b) Angular variation of in-plane slip amplitude at the shank-hole interface at $z = t$, in Models 4C-25 and 4C-27.

**TABLE 8.13. VARIATION OF PARAMETERS THROUGH
THE DEPTH OF THE COUNTERSUNK HOLE**

Model		Installation $\sigma_{\theta\theta}$ (MPa)	Peak $\sigma_{\theta\theta}$ (MPa)	Peak δ_{12} (μm)
4C-24	z=0	0	199.3	51.5
	z=0.5t	0	726.1	47.0
	z=t	0	933.0	55.6
4C-25	z=0	159	198.7	4.3
	z=0.5t	102	336.9	7.6
	z=t	269	544.2	10.6
4C-27	z=0	1	202.8	39.1
	z=0.5t	-1	651.2	37.6
	z=t	29	848.8	45.2

* Installation $\sigma_{\theta\theta}$ values are from the Isotropic response analyses

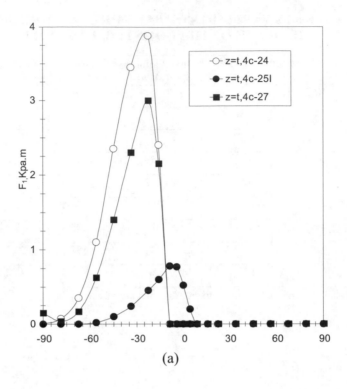

(a)

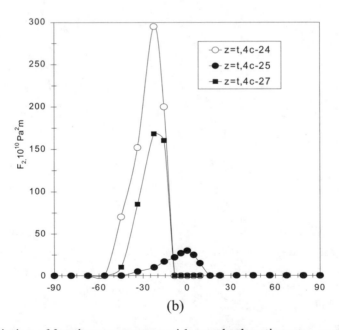

(b)

Figure 8.10. Variation of fretting parameters with angular location at $z = t$, the depth at which the parameters peak, for Models 4C-24, 4C-25 and 4C-27. (a) fretting wear parameter, F_1, and (b) fretting fatigue parameter, F_2.

**TABLE 8.14. SUMMARY OF MECHANISTIC PARAMETERS
INDICATING LOCATIONS OF FRETTING WEAR
AND CRACK INITIATION IN LAP JOINT ASSEMBLIES
WITH COUNTERSUNK RIVETS**

Model		Peak Tangential Stress (MPa)		Peak F_1 (kPa.m)		Peak F_2 (10^{10} Pa^2.m)	
		magnitude (MPa)	$\theta, °$	magnitude (kPa.m)	$\theta, °$	magnitude (10^{10} Pa^2.m)	$\theta, °$
4C-24	z=0	99	45.0, 78.8	1.23	**67.5, 90.0**	7.60	-4.3
	z=0.5t	172	-56.3, -45.0	0.63	-22.5, -15.6	35.00	-15.6
	z=t	309	-45.0	0.98	-22.5	74.35	-22.5
4C-25	z=0	204	56.3, 78.8	0.21	78.8, 90.0	1.03	0.0, 8.7
	z=0.5t	112	4.3, 15.6	0.15	15.6, 22.4	4.70	22.4
	z=t	134	-22.5, -8.7	0.18	-8.7, -4.3	7.50	-4.3
4C-27	z=0	108	45.0, 78.7	0.98	67.5, 90.0	4.63	15.6
	z=0.5t	199	-45.0	0.53	-22.5, -8.7	25.70	-8.7
	z=t	291	-45.0	0.75	-22.5	40.70	-22.5, -15.6

* two angular locations indicates the range over which a peak persists.
- indicates compressive, bulk stresses, and F_2 does not have any significance.

**TABLE 8.15. FEATURES OF THE UPPER (COUNTERSUNK)
AND LOWER (STRAIGHT) PANEL HOLES' DEFORMATION
DUE TO 1% INTERFERENCE**

Model		Installation $\sigma_{\theta\theta}$ (MPa)	Peak $\sigma_{\theta\theta}$ (MPa)	Peak δ_{12} (μm)
5C-3	z=0	0	199.6	36.0
upper hole	z=0.5t	0	505.3	30.0
# 1	z=t	0	609.8	33.5
5C-4	z=0	159	265.5	2.3
upper hole	z=0.5t	102	244.4	2.8
# 1	z=t	269	418.1	3.9
5C-4	z=t	311	446.5	4.0
lower hole	z=1.5t	132	290.1	2.4
# 2	z=2t	75	241.4	1.6

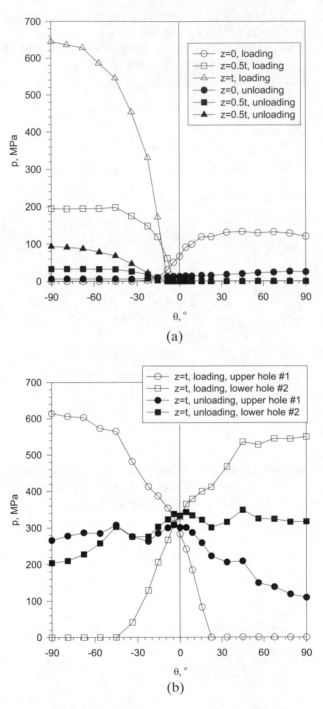

Figure 8.11. (a) Variation of contact pressure at the shank-hole (upper panel hole # 1) inter-face with angular location and depth for Model 5C-3, (b) Variation of contact pressure with angular location in the two fatigue critical holes at depth z = t, where the contact pressures peak, for Model 5C-4.

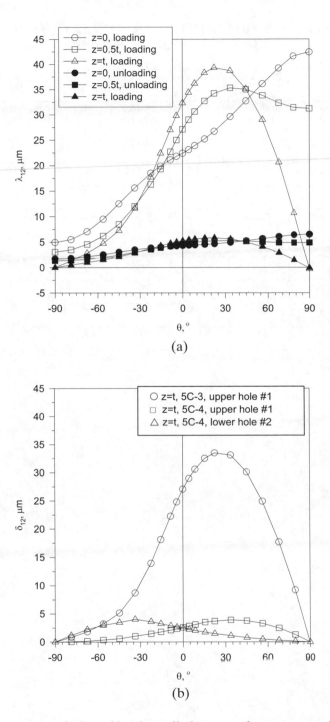

Figure 8.12. (a) Angular variation of in-plane slip between the upper panel hole # 1 and rivet # 1 in Model 5C-3, (b) Angular variation of in-plane slip amplitudes in the fatigue critical holes at depth z = t, where slip amplitudes peak, in Models 5C-3 and 5C-4.

TABLE 8.16. VARIATION OF PARAMETERS THROUGH
THE DEPTH OF THE HOLE

Model		Peak Tangential Stress (MPa)		Peak F_1 (kPa.m)		Peak F_2 (10^{10} Pa^2.m)	
		magnitude (MPa)	θ, °	magnitude (kPa.m)	θ, °	magnitude (10^{10}Pa^2.m)	θ, °
5C-3	z=0	51.4	67.5, 78.8	0.44	78.8, 90.0	3.33	15.6
upper	z=0.5t	78.7	-45.0	0.19	-22.5, -15.6	7.18	-15.6
hole #1	z=t	191.0	-45.0	0.45	-22.5	18.78	-22.5
5C-4	z=0	111.0	78.8	0.05	45, 56.3	0.58	33.7
upper	z=0.5t	48.1	33.7, 45	0.04	33.7	0.78	33.7
hole #1	z=t	40.6	-8.7, -4.3	0.03	0, 4.3	0.98	4.3
5C-4	z=0	44.2	-4.3, 4.3	0.03	-15.6, -8.7	1.33	-15.6
lower	z=0.5t	43.3	-33.7	0.03	-45.0, -33.7	0.48	-33.7
hole #2	z=t	57.9	-35	0.06	78.8, 90.0	1.20	90.0

* two angular locations indicates the range over which a peak persists.

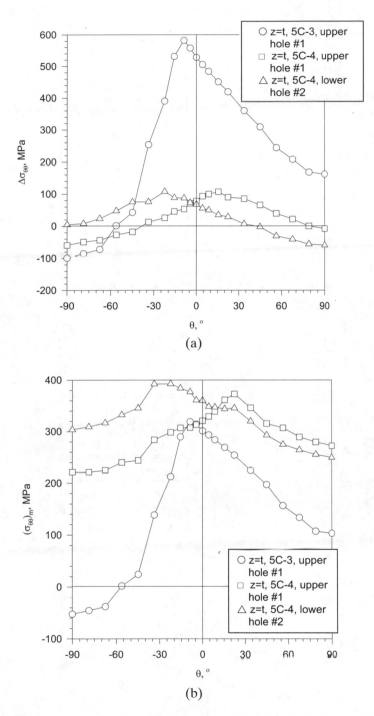

Figure 8.13. Variation of conventional fatigue parameters in the fatigue critical holes in the Models 5C-3 and 5C-4 at z = t. (a) cyclic stress range, (b) mean stress.

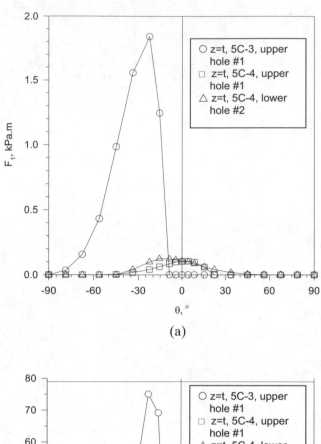

(a)

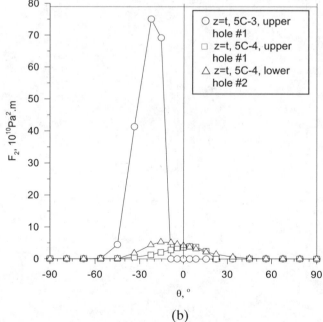

(b)

Figure 8.14. Angular variation of mechanistic fretting parameters at the fretting-fatigue critical hole interfaces in Models 5C-3 and 5C-4 at $z = t$, the depth at which the parameters peak. (a) F_1, (b) F_2.

TABLE 8.17. SUMMARY OF MECHANISTIC PARAMETERS INDICATING LOCATIONS OF FRETTING WEAR AND CRACK INITIATION IN LAP JOINT ASSEMBLIES WITH COUNTERSUNK RIVETS

Deformation Feature	0% Interference		1% Interference	
	Single Rivet-Row 4C-24	Double Rivet-Row 5C-3	Single Rivet-Row 4C-25	Double Rivet-Row 5C-4
Peak p, MPa	810	640	750	610
Peak δ_{12}, μm	58	33	10	4
Peak $\sigma_{\theta\theta}$, MPa	940	610	550	415
Peak $\Delta\sigma_{\theta\theta}$, MPa	900	580	220	90
Peak $\sigma_{m}|_{\theta\theta}$, MPa	480	300	440	360
Peak F_1, kPa.m	3.9	1.8	0.70	0.10
Peak F_2, 10^{10} Pa^2.m	297.4	75.1	30.0	0.12

SINGLE RIVET-ROW LAP JOINTS UNDER BIAXIAL LOADING

9.1 SUMMARY OF CALCULATIONAL MODELS FOR 2-D AND 3-D LAP JOINTS UNDER BIAXIAL LOADING

Table A.8 presents a summary of the calculations done for 2-D and 3-D lap joints under biaxial loading. In the three-dimensional analyses, countersunk and non-countersunk rivet geometries are treated. Interference and clamping stresses generated from rivet installation are not included. Three values of the biaxiality ratio, ϕ, (ratio of the longitudinal to the "hoop" stress) are considered. The finite element models used to obtain these data are described in detail in Chapter 13 [1,17–18]; the material properties for the various models are described in Chapter 15.

9.2 SUMMARY OF CALCULATION RESULTS FOR PINNED CONNECTIONS UNDER BIAXIAL LOADING

Table 12.5 presents the results for biaxial loading in a pinned connection. The data includes joint extension, slip at the pin-panel interface, peak tensile stresses and peak contact pressure [1,18].

Figure 9.1 shows the effects of biaxial loading on the variations of the contact pressure and interfacial slip at the pin-panel interface.

9.3 SUMMARY OF CALCULATION RESULTS FOR SINGLE RIVET-ROW LAP JOINTS UNDER BIAXIAL LOADING

Table 12.6 in Chapter 12 contains the results of calculations for single rivet-row lap joints under biaxial loading. The data includes joint compliance, rivet tilt, total in-plane slip at the

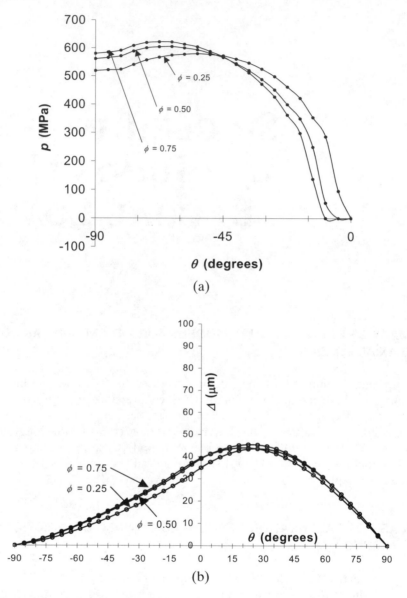

Figure 9.1. Variation of the pin-panel (a) contact pressure, p, and (b) relative slip, Δ, with angular position, θ, in the 2-D pinned connection as a function of the stress biaxiality, ϕ. Owing to symmetry, distributions between $90° \leq \theta \leq 180°$ and $0° \leq \theta \leq 90°$ are identical to each other as are those between $-90° \leq \theta \leq \pm 180°$ and $0° \leq \theta \leq -90°$

rivet-panel interface and the panel-panel interface, total out-of-plane slip, peak tensile stresses and peak contact pressure.

Figures 9.2 and 9.3 show distributions of the contact pressure, p, and in-plane rivet-panel slip, Δ, for the 3-D riveted lap joint and non-countersunk and countersunk rivets, respectively. The plots shown are at a depth of z = 1.53 mm in the upper panel, where the parameters have

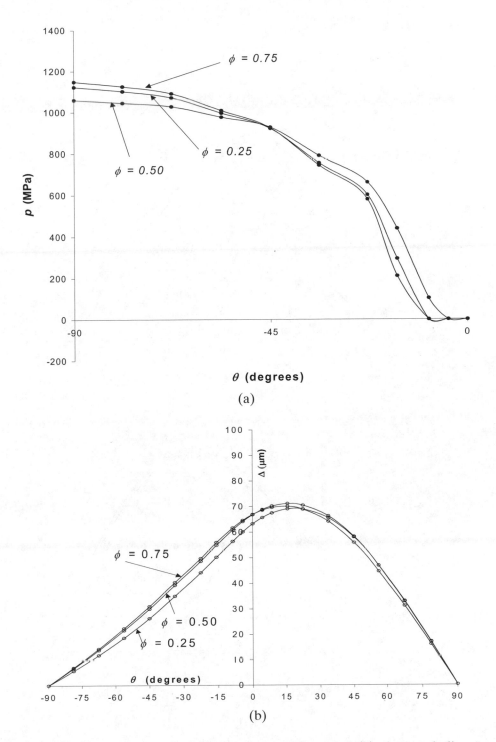

Figure 9.2. Variation of pin-panel (a) contact pressure, p, and (b) pin-panel slip, Δ, with angular position, θ, at $z = 1.53$ mm of the upper panel of the 3-D riveted connection with a non-countersunk rivet.

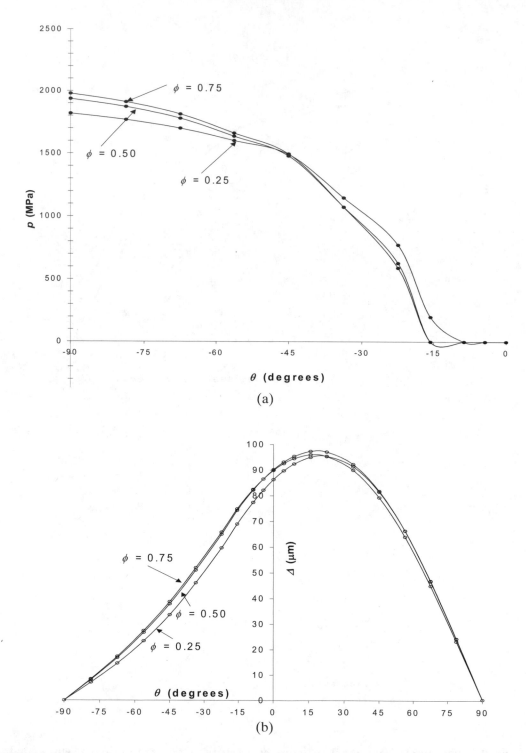

Figure 9.3. Variation of pin-panel (a) contact pressure, p, and (b) pin-panel slip, Δ, with angular position, θ, at $z = 1.53$ mm of the upper panel of the 3-D riveted connection with a countersunk rivet.

TABLE 9.1. INFLUENCE OF STRESS BIAXIALITY ON CALCULATED STRESS CONCENTRATION FACTORS IN THE PANELS OF THE THREE-DIMENSIONAL RIVETED LAP JOINT MODELS

Model	Transverse-to-Longitudinal Stress Ratio	SCF	NPC[*]
4S-16	0.5	6.1	-7.6
4S-17	0.75	5.9	-8.0
4S-18	0.25	5.5	-8.2
4C-17	0.5	8.9	-10.4
4C-18	0.75	8.8	-11.0
4C-19	0.25	8.4	-11.2

[*] NPC = the negative ratio of the peak compressive stress to the nominal tensiile stress based one repeat section.

TABLE 9.2. IMPORTANT STRESSES GENERATED IN THE RIVET

Model	Peak Shank-Hole Compressive Stress (σ_{11}), MPa	Peak Shear Stress (σ_{13}), MPa	Peak Axial Tensile Stress (σ_{33}), MPa
4S-16	-439.5	192.0	341.0
4S-17	-423.4	197.5	343.5
4S-18	-431.5	200.5	344.5
4C-17	-758.5	322.0	428.0
4C-18	-781.0	328.5	435.0
4C-19	-779.0	334.5	438.0

been found to attain peak values and fretting fatigue cracks are known to initiate due to the presence of multiple, superposed stress concentrations arising from the rivet-panel contact and panel bending [17].

Table 9.1 shows the stress concentration factors obtained for the 3-D lap joint models and Table 9.2 lists the maximum values of stresses that characterize the shearing and bending of the rivet.

COMPILATIION OF RESULTS FOR LAP JOINTS WITH SEALANTS AND ADHESIVES

10.1 SUMMARY OF CALCULATIONAL MODELS FOR SINGLE RIVET-ROW LAP JOINTS WITH SEALANTS AND ADHESIVES

Table A.9 presents a summary of the calculations done for single rivet-row lap joints in the presence of sealants. The finite element models used to obtain these data are described in detail in Chapter 13 [6–8,21]; the material properties for the various models are described in Chapter 15.

10.2 SUMMARY OF CALCULATION RESULTS FOR PINNED CONNECTIONS WITH SEALANTS AND ADHESIVES

Table 12.7 contains the results of calculations for single rivet-row sealed lap joints. The data includes joint compliance, rivet tilt, total in-plane slip at the rivet-panel interface and the panel-panel interface, total out-of-plane slip, peak tensile stresses and peak contact pressure.

The term sealant is used to refer to non-linear, rubber-like, low modulus, sticky polymers with pressure dependent shear properties. The term adhesive is used to refer to linear, higher modulus polymers. The joints discussed here are fastened or sealed with a low modulus sealant. The mechanical effects of both sealants and adhesives can be treated with the TALA method (described in Chapter 14) which has been applied to the finite element models presented in Chapter 13. The terms sealant and adhesive are used interchangeably since their effects on lap joints are qualitatively similar.

Figure 10.1 shows the distribution of axial stress (normalized with respect to the applied nominal stress) at the upper panel hole surface for infinitely wide multi-riveted lap joints. The axial stress is plotted as a function of angle (from $-90°$ to 90) and depth (t = 0.0, 0.765 and 1.53 mm).

Figure 10.2 shows the curvature lines on the upper panel of the multi-riveted lap joints with and without sealant.

The peak stresses in the rivet elements are shown in Table 10.1.

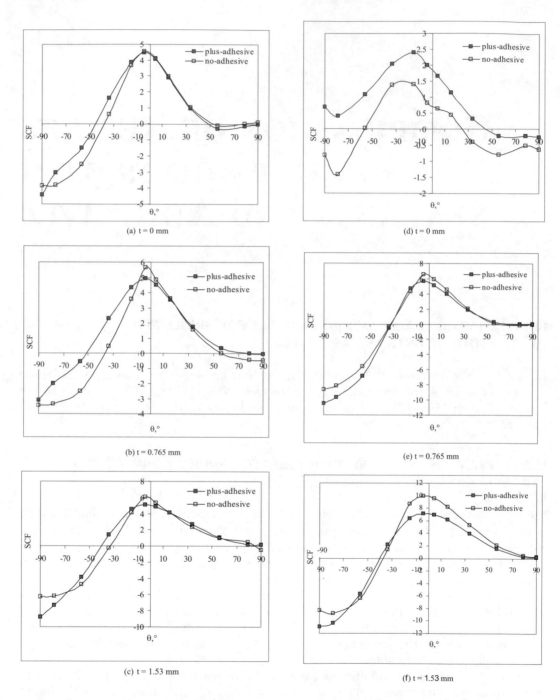

Figure 10.1. Angular variation of normal stress (σ_{11}) around the surface of upper panel hole for infinitely wide multi-riveted lap joints. (a), (b) and (c) are for models 4SA-1 and 4S-6 and (standard single rivet-row plus-adhesive and no-adhesive, respectively) at depth 0, 0.765 and 1.53, respectively. (d), (e) and (f) are for models 4SA-3 and 4C-7 (countersunk single rivet-row plus-adhesive and no-adhesive) at depth 0, 0.765 and 1.53, respectively.

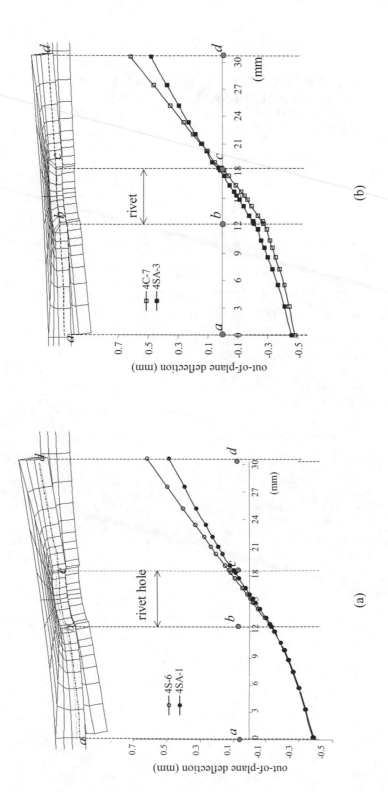

Figure 10.2. Plots of the curvature line on the upper panel of infinitely wide multi-riveted lap joints models (a) 4S-6 (noncountersunk single rivet-row no-adhesive), and 4SA-1 (noncountersunk single rivet-row plus-adhesive), (b) 4C-7 (countersunk single rivet-row no-adhesive), and 4SA-3 (countersunk single rivet-row plus-adhesive).

**TABLE 10.1. STRESSES GENERATED IN RIVETS WITH
AND WITHOUT A SEALANT LAYER**

Model	Peak Shank-Hole Compressive Stress (S11), MPa	Peak Shear Stress (S13), MPa	Peak Axial Tensile Stress (S33), MPa	Peak Axial Compressive Stress (S33), MPa
Standard single rivet-row no-adhesive (4S-7, one-rivet)	-205.8	97.65	127.9	-157.1
Standard single rivet-row plus-adhesive(4SA-2, one-rivet)	-246.0	112.4	86.1	-119.0
Standard single rivet-row no-adhesive(4S-6, multiple-rivet)	-212.1	105.6	127.2	-135.6
Standard single rivet-row plus-adhesive(4SA-1, multiple-rivet)	-252.3	114.8	85.81	-111.3
Countersunk single rivet-row no-adhesive(4C-8, one-rivet)	-332.4	121.0	180.4	-191.2
Countersunk single rivet-row plus-adhesive (4CA-4, one-rivet)	-364.2	130.0	87.28	-128.9
Countersunk single rivet-row no-adhesive (4C-7, multiple-rivet)	-314.4	123.7	181.9	-190.8
Countersunk single rivet-row plus-adhesive(4CA-3, multiple-rivet)	-373.6	133.1	87.77	-107.8

11

COMPILATION OF
RESULTS FOR RIVETS

11.1 SUMMARY OF CALCULATIONAL MODELS FOR RIVETS

Table A.4 presents a summary of the calculations done to evaluate the effects of rivet installation. The finite element models used to obtain these data are described in detail in Chapter 13; the material properties for the various models are described in Chapter 15.

11.2 SUMMARY OF CALCULATIONAL RESULTS FOR RIVETS

Table 11.1 summarizes important stresses generated in the rivet shank.

Table 11.2 shows the variation in the distribution of in- and out-of-plane load transmission with the number of rivet-rows. The peak rivet shear stress, σ_{13}, indicative of in-plane transmission, and the peak rivet axial stress, σ_{33}, indicative of out-of-plane transmission, are presented. Two parameters which are easily deduced and indicate the relative proportions of in- and out-of-plane load transmission are listed.

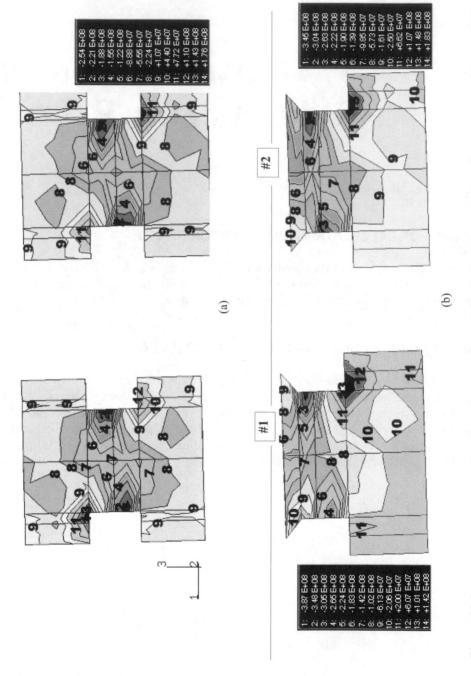

(a)

(b)

Figure 11.1. σ_{11} stresses generated in the shanks of rivets # 1 and # 2 due to in-plane load transfer. (a) model 5S-1 (b) model 5C-1, σ_{nom} = 125 MPa.

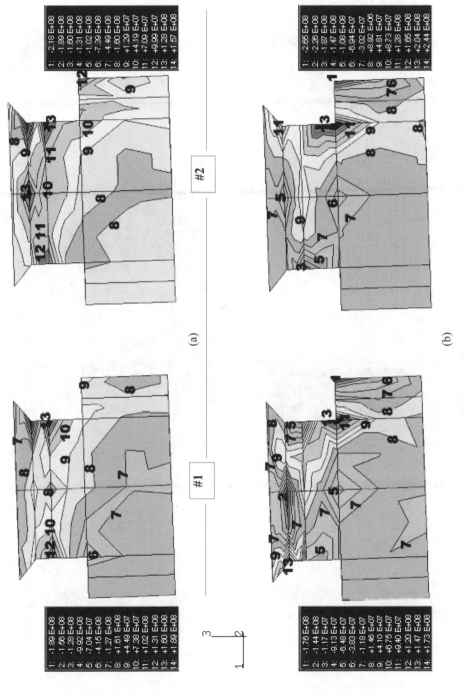

Figure 11.2. Stresses generated in the shanks of rivets #1 and # 2 in model 5C-1: (a) Shear stresses (σ_{13}) and (b) Axial stresses (σ_{33}), $\sigma_{\text{nom}} = 125$ MPa.

163

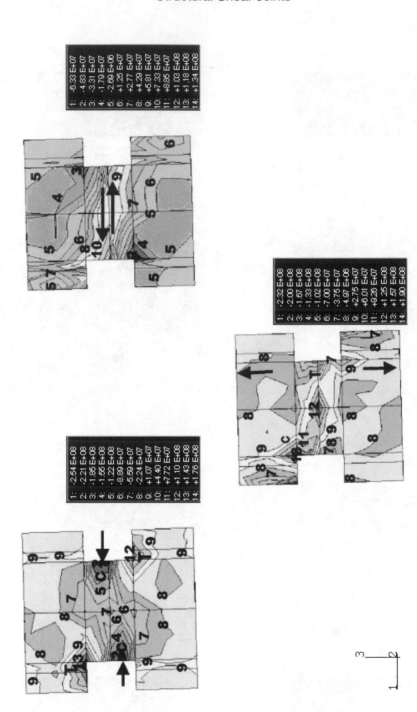

Figure 11.3. σ_{11}, σ_{13} and σ_{33} stresses generated in the shank of rivet #1 due to in-plane and out-of-plane load transfer. The letters C and T are used to denote compressive and tensile regions, respectively (non-countersunk case), σ_{nom} = 125 MPa.

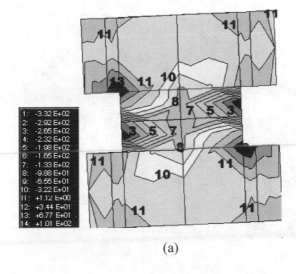

(a)

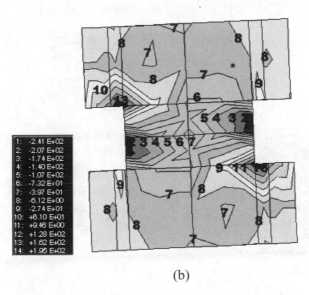

(b)

Figure 11.4. Contour plots of σ_{11} (normal stress, in-plane direction) generated in rivets for infinitely wide multi-riveted lap joints (a) model 4SA-1 (standard rivet plus-adhesive) and (b) model 4S-6 (standard rivet no-adhesive), $\sigma_{nom} = 65$ MPa.

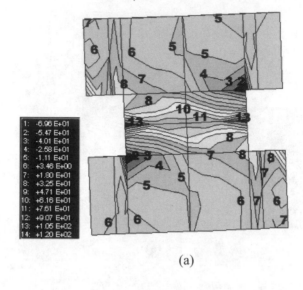

(a)

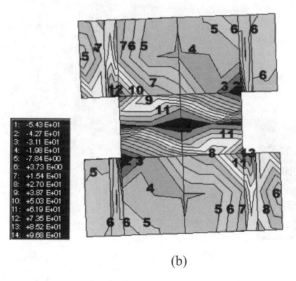

(b)

Figure 11.5. Contour plots of σ_{13} (shear stress) generated in rivets for infinitely wide multi-riveted lap joints (a) model 4SA-1 (standard rivet plus-adhesive) and (b) model 4S-6 (standard rivet no-adhesive), σ_{nom} = 65 MPa.

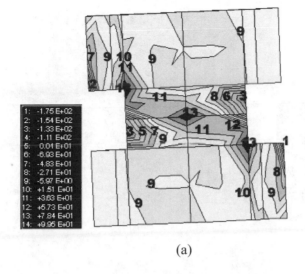

(a)

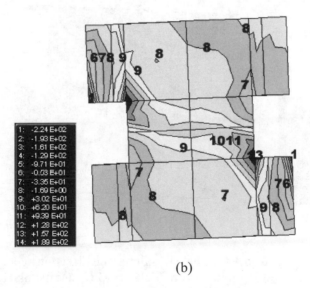

(b)

Figure 11.6. Contour plots of σ_{33} (axial stress, out-of-plane direction) generated in rivets for infinitely wide multi-riveted lap joints (a) model 4SA-1 (standard rivet plus-adhesive) and (b) model 4S-6 (standard rivet no-adhesive), $\sigma_{nom} = 65$ MPa.

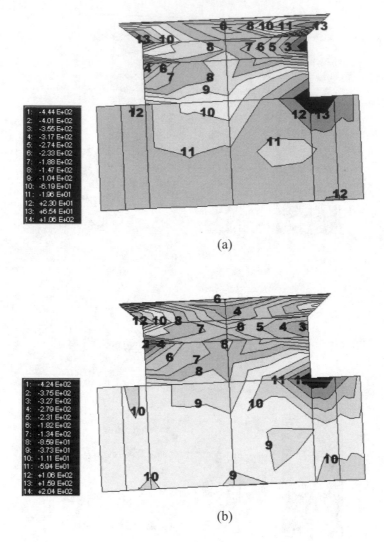

(a)

(b)

Figure 11.7. Contour plots of σ_{11} (normal stress, in-plane direction) generated in rivets for infinitely wide multi-riveted lap joints (a) model 4CA-3 (countersunk rivet plus-adhesive) and (b) model 4C-7 (countersunk rivet no-adhesive), $\sigma_{nom} = 65$ MPa.

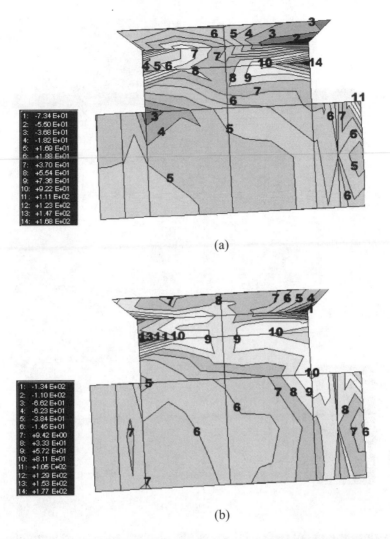

(a)

(b)

Figure 11.8. Contour plots of σ_{13} (shear stress) generated in rivets for infinitely wide multi-riveted lap joints (a) model 4CA-3 (countersunk rivet plus-adhesive) and (b) model 4C-7 (countersunk rivet no-adhesive), $\sigma_{nom} = 60$ MPa.

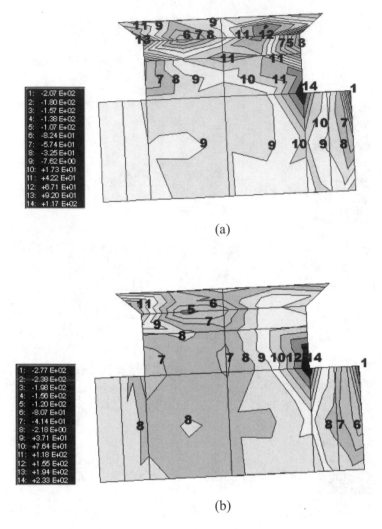

(a)

(b)

Figure 11.9. Contour plots of σ_{33} (axial stress, out-of-plane direction) generated in rivets for infinitely wide multi-riveted lap joints (a) model 4CA-3 (countersunk rivet plus-adhesive) and (b) model 4C-7 (countersunk rivet no-adhesive), σ_{nom} = 60 MPa.

TABLE 11.1. IMPORTANT STRESSES GENERATED IN THE RIVET SHANK OF DOUBLE ROW LAP JOINTS

Model	Peak Shank-Hole Compressive Stress (σ_{11}), MPa		Peak Shear Stress (σ_{13}), MPa		Peak Axial Tensile Stress (σ_{33}), MPa		Peak Axial Compressive Stress (σ_{33}), MPa	
	rivet # 1	rivet # 2	rivet # 1	rivet # 2	rivet # 1	rivet # 2	rivet # 1	rivet # 2
5S-1	-240	-240	126	126	174	174	-95	-95
5C-1	-353	-321	175	143	160	224	-115	-164
5C-2	-338	-335	168	133	200	266	-116	-157

TABLE 11.2. COMPARISONS OF PROPORTIONS OF SHEAR AND ROTATION IN RIVETS IN SINGLE AND DOUBLE ROW LAP JOINT MODELS.

Model	Rivet #	Single Rivet-Row (MPa)		Double Rivet-Row (MPa)		σ_{13}/σ_{33}		$\dfrac{(\sigma_{13})}{((\sigma_{13})^2+(\sigma_{33})^2)^{0.5}}$	
		σ_{13}	σ_{33}	σ_{13}	σ_{33}	Single Row	Double Row	Single Row	Double Row
4S-1 or	1	239	309	126	174	0.77	0.72	0.61	0.59
5S-1	2	-	-	126	174	-	0.72	-	0.59
4C-1 or	1	325	418	175	160	0.78	1.09	0.61	0.74
5C-1	2	-	-	143	224	-	0.64	-	0.54
4C-2	1	388	489	168	200	0.79	0.84	0.62	0.64
5C-2	2	-	-	133	266	-	0.5	-	0.45

DATA TABULATIONS

The following tables present summaries of the results of many sof the finite element calculations described in this book. The model numbers refer to the joint configurations and fastener types (see Introduction for code keys and definitions of computational models). Details of each model number may be found in the tables in Appendix A. The finite element models used to obtain these data are described in detail in Chapter 13; the material properties for the various models are described in Chapter 15.

The effects of rivet geometry (countersinking), rivet material and interfacial friction coefficient are included. Interference and lateral clamping are not treated. Load transfer through the joint, the joint compliance, rivet tilt, the local slips at rivet-panel and panel-panel interfaces, contact pressures and local stresses are presented.

The *excess compliance* is defined as $C = C' - C''$. C', the assembly compliance, is the ratio of the net extension, δ, to the load per repeat distance, P. C'' is the compliance of a continuous panel having the same length, L, as the model joint (length of two panels minus the overlap) and the same cross sectional area, A, and elastic modulus, E, as a single panel. The excess compliance is related to the displacement, bending and deformation around the rivet.

The *rivet tilt* is defined by the change in the slope of the line AB, illustrated in Fig. 12.1, which bisects the top of the rivet heads.

The *stress concentration factor* (SCF) is defined as the ratio of the peak tensile stress to the nominal applied stress, based on one repeat section. The angular location convention is described in Figure 12.2.

12.1 SUMMARY CALCULATION RESULTS FOR SINGLE RIVET-ROW LAP JOINTS

The data presented in Table 12.1 are valid for elastic, and elastic-plastic single rivet-row, aluminum alloy lap joints with $C'' = 93.4$ m/GN.

12.2 SUMMARY CALCULATION RESULTS FOR DOUBLE RIVET-ROW LAP JOINTS

The data presented in Table 12.2 are valid for elastic, double rivet-row, aluminum alloy lap joints with standard and countersunk aluminum and steel rivets and $C'' = 102.71$ m/GN. They are compared with similar analyses for single rivet-row joints.

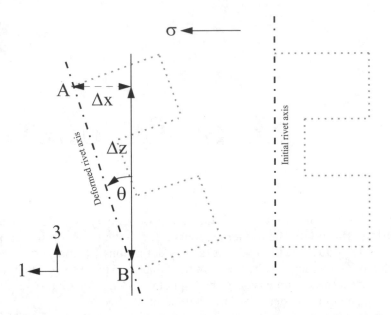

Figure 12.1. Schematic showing rivet displacements used to calculate tilt values reported in Table 12.2. A and B represent the coordinates of the end points of the line bisecting the rivet heads.

12.3 SUMMARY CALCULATION RESULTS FOR RIVET INSTALLATION RESIDUAL STRESSES

Tables 12.3 and 12.4 present some of the effects of rivet installation for a single rivet-row aluminum lap joint for standard and countersunk rivet geometries, respectively. Contact pressures, slip amplitudes, bulk (circumferential) stresses, and the local cyclic stress range and mean stress adjacent to the rivet hole are evaluated during the first load cycle, after the application of interference and/or clamping with $C'' = 93.4$ m/GN. They are compared with similar analyses for single rivet-row joints.

12.4 SUMMARY CALCULATION RESULTS FOR SINGLE RIVET-ROW LAP JOINTS UNDER BIAXIAL LOADING

Table 12.5 lists values of panel deformation characteristics of the two-dimensional pinned connection for the three values of ϕ considered in the study, including the excess joint extension, δ.

Table 12.6 summarizes the effects of load biaxiality, ϕ, on the net longitudinal joint extension, excess longitudinal joint extension, rivet tilt, and the panel stresses, slips and pressures in 3-D connections with non-countersunk rivets (models 4S-16, 4S-17 and 4S-18) and countersunk rivets (models 4C-17, 4C-18 and 4C-19).

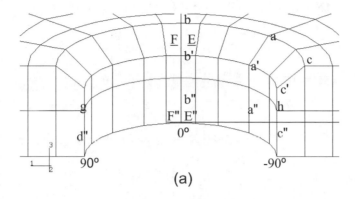

(a)

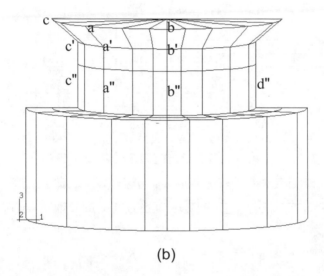

(b)

Figure 12.2. Details of the mesh showing rivet type B and the panels. Values of peak stresses in elements $\underline{E}$, $\underline{E}$", $\underline{F}$ and $\underline{F}$" and slips at nodes a-d and g-h are presented in the tables in this chapter. Slips at nodes a-d are calculated by subtracting the displacement of the rivet from the displacement of the sheet.

12.5 SUMMARY CALCULATION RESULTS FOR SINGLE RIVET-ROW LAP JOINTS WITH SEALANT

Table 12.7 shows the influence of the thin sealant layer on the distortion of the riveted lap joint. The local mechanical parameters: excess compliance, rivet tilt, in-plane and out-of-plane slip at certain interesting points (a, a', a", b, b', b", c, c', c", d", g, and h, in Figure 12.2), peak tensile stress in panel hole, peak contact pressure, panel-panel separation (observing from points m and n in Figure 12.4), stresses in the panel and in the rivet, and panel bending

TABLE 12.1. SINGLE RIVET-ROW LAP JOINT DISTORTION FEATURES

Model Feature	Location*	4S-1 elastic, μ=0.2	4S-2 elastic, μ=0.2	4S-3 plastic, μ=0.2	4C-1 elastic, μ=0.2	4C-2 elastic, μ=0.2	4C-3 plastic, μ=0.2	4C-4 plastic, μ=0.4	4C-5 elastic, μ=0.2	4C-6 elastic, μ=0.2
Joint Compliance (m/GN)	—	27.46	23.48	26.64	34.4	28.3	43.6	32.8	36.74	41.18
Rivet Tilt, degrees	—	3.6	3.67	3.56	4.24	3.97	4.41	4.35	4.46	4.13
Total in-plane slip at rivet-panel interface (μm)	a	16.47	8.09	18.03	32.2	24.2	65.6	31.03	36.3	66.27
	a'	16.14	7.3	16.63	25.38	18.2	52.6	15.29	37.1	63.38
	a"	-46.39	-36.46	-49.4	-46.6	-38.7	-47.9	-39.6	-39.4	-33.29
	b	42.2	30.6	47.7	59	47.5	108.6	62.7	63.1	110.0
	b'	45.0	32.6	51.3	67.1	52.5	125.8	77.3	81.6	121.7
	b"	-45.0	-32.6	-51.3	-45.1	-36.7	-50.5	-39.3	-39.2	-34.0
Total in-plane slip at panel-panel interface (μm)	g	91.3	74.1	96.3	147.3	117.7	213.0	160.1	169.6	202.1
	h	91.4	74.1	96.3	97.2	68.8	100.0	88.4	89.0	89.5
Total out-of-plane slip (μm)	c	-5.9	-2.3	-5.9	-17.7	-11.6	-53.1	-13.4	-44.77	-65.7
	c'	-2.7	-1	-2.4	-14.6	-9.17	-44.84	-7.22	-41.4	-60.5
	c"	8.7	9.4	9.1	7.53	7.0	1.45	10.66	-10.1	-20.14
	d"	2.6	1.0	2.4	5.6	4.2	5.4	2.2	3.4	2.0
Peak tensile stress in elements adjacent to rivet hole (MPa)	E	766.8	765.6	549.4	1220.9	1097.4	633.8	614.4	1306.9	1512.4
	E"	706.6	667.5	517.3	690.2	614.4	549.3	522.1	641.0	499.4
	F	706.8	667.5	516.8	1180.6	1059.8	593.6	592.7	1242.3	1357.3
	F"	767.1	767.25	549.8	738.9	671.7	563.5	551.5	707.8	597.4
Peak Contact Pressure (MPa)	shank- hole (A)**	779.2	641.6	744.24	1052.5	905.5	1141.5	963.7	1349.4	1953.4
	panel-panel (B)**	350.7	347.2	364.4	334.9	378.2	445.6	324.1	822.7	1252.0
Contact Pressure (MPa)	panel-rivet head (C)**	228.6	246.4	244.3	283.6	347.4	306.9	329.9	256.9	206.7

* see Figure 12.2
** shown in Figure 12.3

TABLE 12.2. LAP JOIN DISTORTION FEATURES FOR THREE-DIMENSIONAL SINGLE AND DOUBLE RIVET-ROW ASSEMBLIES WITH NON-COUNTERSUNK AND COUNTERSUNK RIVETS

Model		Non-countersunk Aluminum Rivets			Countersunk Aluminum Rivets			Countersunk Steel Rivets		
Feature	Location*	4S-1 Single row	5S-1 Double row		4C-1 Single row	5C-1 Double row		4C-2 Single row	5C-2 Double row	
			Rivet #1	Rivet #2		Rivet #1	Rivet #2		Rivet #1	Rivet #2
Excess Compliance (m/GN)	—	27.46	7.45		34.4	10.32		28.3	7.57	
Rivet Tilt, degrees	—	3.6	1.68	4.24	4.24	1.47	1.87	3.97	1.87	1.79
Total in-plane slip at rivet-panel interface (μm)	A	16.5	9.6	32.2	32.2	16.2	17.0	24.2	13.2	12.4
	a'	16.1	9.2	25.4	25.4	13.2	18.8	18.2	10.3	9.7
	A"	-46.4	-23.7	-46.6	-46.6	-23.8	-30.5	-38.7	-19.3	-26.3
	B	42.2	26.4	59	59	32.3	30.8	47.5	27.8	24.1
	B'	45.0	27.7	67.1	67.1	37.2	33.0	52.5	31.2	25.6
	B"	-45.0	-23.6	-45.1	-45.1	-23.5	-27.5	-36.7	-18.9	-23.2
Total in-plane slip at panel-panel interface (μm)	G	91.3	58.9	147.3	147.3	85.0	63.6	117.7	71.9	49.6
	H	91.4	44.1	97.2	97.2	46.1	60.6	68.8	33.0	46.3
Total out-of-plane slip (μm)	c	-5.9	-1.9	-17.7	-17.7	-6.6	-9.2	-11.6	-4.7	-6.3
	c'	-2.7	-1.3	-14.6	-14.6	-5.7	-8.6	-9.17	-4.2	-6.0
	c"	8.7	0.5	7.53	7.53	-1.0	2.2	7.0	-0.2	2.4
	d"	2.6	2.3	5.6	5.6	4.4	3.7	4.2	3.5	3.5
Peak tensile stress in elements adjacent to rivet hole (MPa)	E	766.8	549.7	1220.9	1220.9	758.6	558.2	1097.4	740.4	524.8
	E"	706.6	351.6	690.2	690.2	348.6	497.0	614.4	317.1	465.3
	F	706.8	517.6	1180.6	1180.6	729.8	514.0	1059.8	716.8	493.7
	F"	767.1	384.4	738.9	738.9	372.2	525.9	671.7	344.4	499.0
Peak Conta Contact Pressure (MPa)	Shank-hole (A)**	779.2	471.7	1052.5	1052.5	719.1	496.5	905.5	561.2	478.1
	Panel-panel (B)**	350.7	188.3	334.9	334.9	125.1	177.9	378.2	115.0	213.5
Contact Pressure (MPa)	Panel-rivet head (C)**	228.6	121.0	283.6	283.6	78.8	133.3	347.4	106.9	174.4

*Locations identified in Fig. 12.2.
**Contact field with which pressure is associated (Fig. 12.3)

**TABLE 12.3. DISTORTION FEATURES FOR LAP
JOINTS WITH NON-COUNTERSUNK RIVETS,
UNDER MAXIMUM LOAD, σ_{max} = 100 MPa**

Model		4S-19	4S-20	4S-22	4S-23
Feature	Location *				
Excess Compliance (m/ GN)	—	25.3	15.4	20.7	13.9
Rivet Tilt, degrees	—	3.21	3.15	3.11	3.04
Total in-plane slip at rivet- panel interface (μm)	a	8.84	0.75	5.73	0.07
	a'	7.92	1.20	7.14	0.85
	a"	-32.3	-5.65	-23.91	-4.60
	b	28.7	3.0	18.7	1.70
	b'	30.7	4.1	22.7	3.60
	b"	-30.7	-4.1	-22.9	-3.10
Total in-plane slip at panel- panel interface (μm)	g	65.2	11.6	48.4	5.1
	h	65.2	11.6	48.7	9.7
Total out-of-plane slip (μm)	c	-2.7	-12.9	-14.3	-20.6
	c'	-1.1	-3.0	-12.7	-6.6
	c"	7.8	6.7	-3.0	3.4
	d"	1.1	3.0	-2.2	1.7
Peak tensile stress in elements adjacent to rivet hole (MPa)	e	531.5	373.1	413.2	336.3
	e"	495.8	433.8	374.4	414.4
	f	495.4	434.5	367.9	352.3
	f"	532.4	374.4	420.6	352.9
Peak Contact Pressure (MPa)	shank- hole (A)*	561.4	558.0	453.1	570.0
	panel- panel (B)*	290.2	241.4	340.5	213.2
Contact Pressure (MPa)	panel- rivet head (C)*	182.4	168.9	144.3	170.1

* denotes the contact field, shown in Fig. 9.1, with which the value is associated.

TABLE 12.4. DISTORTION FEATURES FOR LAP JOINTS WITH COUNTERSUNK RIVETS, UNDER MAXIMUM LOAD, $\sigma_{max} = 100$ MPa

Model		4C-24	4C-25	4C-27
Feature	Location *			
Excess Compliance (m/GN)	—	32.9	17.6	30.4
Rivet Tilt, degrees	—	3.75	3.38	3.63
Total in-plane slip at rivet-panel interface (μm)	a	20.36	1.56	17.25
	a'	10.96	1.06	9.05
	a"	-32.31	-5.80	-29.00
	b	42.60	5.10	37.9
	b'	52.50	5.40	46.6
	b"	-30.60	-4.10	-27.5
Total in-plane slip at panel-panel interface (μm)	g	116.8	26.4	108.7
	h	70.4	12.0	60.9
Total out-of-plane slip (μm)	c	-8.4	-20.5	-7.2
	c'	-5.3	-12.7	-2.0
	c"	8.3	6.8	15.4
	d"	2.3	3.5	15.4
Peak tensile stress in elements adjacent to rivet hole (MPa)	e	581.8	432.6	558.4
	e"	491.9	405.2	424.2
	f	572.1	499.2	577.4
	f"	530.1	379.5	490.1
Peak Contact Pressure (MPa)	shank-hole (A)*	766.8	669.1	768.9
	panel-panel (B)*	361.1	396.5	372.8
Contact Pressure (MPa)	panel-rivet head (C)*	204.6	182.4	187.4

* denotes the contact field, shown in Fig. 9.1, with which the value is associated.

TABLE 12.5. DISTORTION FEATURES OF AN ELASTIC, TWO-DIMENSIONAL PINNED CONNECTION SUBJECTED TO A LONGITUDINAL STRESS OF σ_L = 125 MPa AND THREE VALUES OF TRANSVERSE STRESS, σ_T. BOTH THE PIN AND PLATE ARE MADE OF ALUMINUM AND μ = 0.2

2D Model			
	3P-22	**3P-23**	**3P-24**
Feature	(ϕ = 0.25)	(ϕ = 0.50)	(ϕ = 0.75)
Transverse far-field plate boundary displacement* (μm)	0.0•	±6.85	±13.66•
Transverse, far-field plate boundary stress (MPa)	31.3	62.5•	93.8
δ, excess longitudinal joint extension (μm)	50.4	55.4	53.1
Angular location (range) over which panel-pin contact is lost (°)	[0,180]	[-4.5,+184.5]	[-9,+189]
Peak tensile stress along the longitudinal direction in elements adjacent to hole (MPa)	802.2	794.7	730.8
Angular location (range) of peak panel tensile stress (°)	[0]	[-4.5]	[-14,-9]
Pin-panel contact pressure, p, at θ = -90° (MPa)	522.2	573.0	593.2
Slip, Δ, at pin-panel interface at θ = 0° (μm)	34.9	39.0	39.3
Slip, Δ, at pin-panel interface at θ = -45° (μm)	6.35	7.66	7.91

• applied transverse boundary condition.

TABLE 12.6. DISTORTION FEATURES OF ELASTIC, THREE-DIMENSIONAL RIVETED LAP JOINTS

3D Model Feature	Location *	4S-16	4S-17	4S-18	4C-17	4C-18	4C-19
Net longitudinal joint extension (μm)	—	728.1	703.4	664.7	796.6	774.0	734.3
Transverse far-field plate boundary stress (MPa)	—	31.2	62.5	93.7	31.2	62.5	93.7
Rivet tilt (°)	—	3.55	3.52	3.54	4.41	4.11	4.42
Total in-plane slip at rivet-panel interface (μm)	a	24.1	25.6	28.4	32.5	31.6	37.0
	a'	25.3	28.8	29.8	28.0	31.5	32.3
	a"	-53.7	-53.8	-51.4	-75.7	-76.8	-73.7
	b	52.1	55.8	56.0	55.0	57.8	57.7
	b'	57.1	60.8	61.1	68.4	72.4	72.5
	b"	-57.1	-60.8	-61.1	-86.3	-91.0	-90.7
Total in-plane slip at panel-panel interface (μm)	g	97.0	99.9	98.1	150.9	155.3	152.7
	h	96.9	100.0	98.1	134.9	139.9	137.6
Total out-of-plane slip (μm)	c	-27.6	-29.4	-30.2	-44.4	-43.0	-47.2
	c'	-24.8	-26.4	-27.0	-25.52	-26.6	-27.3
	c"	-4.2	-4.5	-4.9	1.16	1.2	0.6
	d"	24.8	26.3	27.0	82.3	86.9	87.2
Peak tensile stress in elements adjacent to rivet hole (MPa)	e	767.6	736.4	690.2	1117.6	1097.6	1051.2
	e"	642.1	618.5	572.4	706.5	706.0	657.2
	f	642.3	618.5	572.5	1071.9	1063.1	1012.3
	f"	767.5	737.1	690.8	884.3	870.3	825.6
Peak Contact Pressure (MPa)	shank-hole (A)**	947.7	1000.1	1022.3	1299.7	1373.0	1395.6
	panel-panel (B)**	558.6	576.3	580.6	532.9	546.1	540.0
Contact Pressure (MPa)	panel-rivet head (C)**	210.2	208.0	198.6	285.0	284.9	290.8

*Locations identified in Fig. 6.2.
**Contact field with which value is associated.

TABLE 12.7. INFLUENCE OF A THIN SEALANT LAYER ON THE DISTORTION OF INFINITELY WIDE AND FINITE WIDTH SINGLE-RIVETED LAP JOINTS (AN A IN THE MODEL NUMBER INDICATES THE PRESENCE OF ADHESIVE)

Features	Location	standard riveted lap joints				countersunk riveted lap joints			
		multiple-rivet		one-rivet		multiple-rivet		one-rivet	
		4SA-1	4S-6	4SA-2	4S-7	4SA-3	4C-7	4SA-4	4C-8
Excess Compliance (m/GN)	-	38.25	33.27	44.93	39.95	39.88	35.18	46.50	41.86
Rivet Tilt, Degree	-	2.46	2.52	2.47	2.52	2.74	2.82	2.77	2.84
Total in-plane slip at rivet-panel interface (mm)	a	12.26	8.65	11.81	8.26	17.75	17.24	17.24	16.50
	a'	15.56	8.29	15.02	7.91	17.84	13.02	17.24	12.35
	a"	-29.00	-23.37	-30.41	-24.68	-29.27	-23.78	-30.50	-25.01
	b	26.60	21.50	26.60	21.40	32.50	30.20	32.90	30.30
	b'	31.70	22.90	31.60	22.50	38.90	34.10	38.80	33.80
	b"	-31.70	-22.90	-31.60	-22.50	-31.40	-23.20	-31.20	-22.90
Total in-plane slip at panel-panel interface (mm)	g	67.80	46.60	68.80	47.50	84.70	76.80	84.70	78.10
	h	67.80	46.60	68.70	47.40	72.50	49.50	72.50	50.20
Total out-of-plane slip (mm)	c	-29.72	-3.56	-29.37	-3.48	-27.50	-12.96	-27.00	-12.64
	c'	-31.20	-1.24	-30.83	-1.18	-20.93	-7.00	-20.60	-6.82
	c"	10.21	2.89	10.35	2.95	15.43	1.11	15.45	1.16
	d"	31.20	1.30	31.20	1.20	68.40	2.90	67.10	2.90
Peak tensile stress in elements adjacent to rivet hole (MPa)	e	320.2	382.2	340.1	406.6	443.9	622.6	478.0	657.1
	e"	309.5	353.6	321.1	380.4	286.7	334.2	306.6	352.4
	f	309.5	353.6	321.1	380.4	443.9	615.8	475.5	655.3
	f"	320.2	382.2	340.1	406.6	295.5	362.7	311.9	379.3
SCF	-	5.1	6.1	5.4	6.5	7.1	10.0	7.6	10.5
Peak Contact Pressure (MPa)	Shank-hole	506.3	436.7	492.8	423.6	842.7	660.0	819.6	637.9
	Panel-panel	11.0	434.1	10.9	420.9	9.0	409.3	8.9	410.1
	Panel-rivet head	245.1	244.7	245.5	244.5	284.8	545.3	286.6	542.3
Overlapping End Openning (mm)	m-n	0.11	0.24	0.11	0.24	0.12	0.29	0.13	0.29

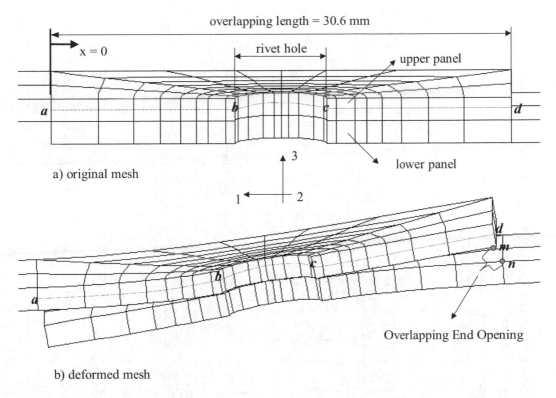

a) original mesh

b) deformed mesh

Figure 12.3. Illustrations of overlapping end opening and the curvature line a-b-c-d on the upper panel.

(observed from curvature line a-b-c-d in Figure 12.3) for each model are examined and compared at the same tensile loading. $C'' = 93.4$ m/GN for these models.

12.6 SUMMARY CALCULATION RESULTS FOR VARIATION OF PANEL THICKNESS IN WIDE SINGLE RIVET-ROW JOINTS

Tables 12.8 and 12.9 show the affect of varying the panel thickness on the various mechanical parameters. The models employed are identical to (4S-1) and (4C-1) except for the panel thicknesses.

12.7 SUMMARY CALCULATION RESULTS FOR JOINT EXCESS COMPLIANCE AND RIVET TILT

Table 12.10 presents data for excess compliance and rivet tilt for several clamped and unclamped lap and butt joints.

TABLE 12.8. THE INFLUENCE OF PANEL THICKNESS ON VARIOUS PANEL PARAMETERS FOR STANDARD HEAD MODEL

Model 4S-1

Feature	Location	Panel Thickness										
		0.77	0.92	1.07	1.15	1.22	1.38	1.53	1.91	2.30	2.68	3.06
Excess Compliance (m/GN)	--	38.68	34.79	32.26	32.12	30.54	29.35	28.80	27.31	26.93	26.24	25.80
Rivet Tilt (degree)	--	3.21	3.29	3.36	3.42	3.41	3.56	3.63	3.72	3.81	3.85	3.87
Total in-plane slip at rivet-panel interface (mm)	a	15.77	16.03	16.24	16.23	16.41	16.62	16.83	17.56	17.74	17.67	17.08
	a'	14.68	14.97	15.26	15.38	15.54	15.84	16.15	16.93	17.48	18.01	18.31
	a"	-45.56	-45.58	-45.74	-47.07	-46.05	-46.43	-47.63	-48.73	-51.35	-52.02	-53.32
	b	42.60	42.40	42.60	42.90	42.70	42.70	43.00	42.50	41.00	38.90	36.10
	b'	42.80	43.20	43.80	44.10	44.30	44.90	45.60	47.20	48.30	49.10	49.50
	b"	-42.90	-43.20	-43.80	-45.70	-44.30	-45.00	-46.50	-47.70	-50.80	-51.30	-51.20
Total in-plane slip at rivet-panel interface (mm)	g	76.40	78.20	81.00	83.60	84.60	88.80	94.40	107.50	121.30	135.20	148.60
	h	76.10	78.20	81.00	84.40	84.50	88.70	94.30	107.30	123.10	136.50	149.70
Total out-of-plane slip (mm)	c	-4.60	-4.80	-5.10	-5.30	-5.50	-5.90	-6.40	-7.90	-9.30	-10.67	-11.86
	c'	-0.90	-0.90	-1.00	-1.10	-1.20	-1.40	-1.80	-2.80	-4.20	-6.10	-8.64
	c"	0.60	2.20	3.70	4.40	5.20	6.50	7.20	7.40	6.00	2.70	-1.92
	d"	-1.00	0.90	1.00	0.00	1.10	1.40	1.70	2.30	3.10	4.50	6.40
Peak tensile stress in elements adjacent to rivet hole (MPa)	e	734.5	734.2	734.2	734.4	737.4	745.0	761.7	814.3	881.8	968.1	1056.6
	e"	691.8	682.7	681.7	692.5	685.3	693.5	715.0	763.8	842.4	923.3	1005.2
	f	694.6	682.8	681.7	688.3	685.3	693.4	716.2	768.6	831.9	913.6	995.7
	f"	742.8	733.9	734.0	746.8	737.4	744.9	767.3	813.2	893.0	978.6	1066.5
Peak Contact Pressure (MPa)	Shank-hole (A)	674.1	721.9	759.4	768.2	793.6	827.4	859.3	935.0	1015.7	1108.4	1218.9
	Panel-Panel (B)	165.1	309.3	456.8	529.6	609.8	762.5	833.6	825.1	842.1	661.0	494.9
Contact Pressure (MPa)	Panel-Rivet head (C)	313.9	311.9	308.0	301.3	324.0	392.4	457.7	616.4	768.0	896.4	999.1

184

TABLE 12.9. THE INFLUENCE OF PANEL THICKNESS ON VARIOUS PANEL PARAMETERS FOR COUNTERSUNK MODEL

Model 4C-1 Feature	Location	0.77	0.92	1.07	1.15	1.53	1.91	2.30	2.68	3.06
Excess Compliance (m/GN)	--	39.77	36.10	33.78	32.89	30.82	29.09	28.07	27.45	27.08
Rivet Tilt (degree)		4.27	4.26	4.14	4.20	4.31	4.38	4.43	4.46	4.40
Total in-plane slip at rivet-panel interface (mm)	a	44.83	41.35	39.16	38.15	36.02	33.37	30.22	26.48	22.26
	a'	33.00	30.13	28.68	28.09	30.87	30.64	30.76	30.91	30.88
	a"	-44.86	-59.63	-46.11	-46.42	-48.48	-49.32	-50.04	-50.73	-51.30
	b	83.20	76.00	71.30	69.20	63.10	56.20	49.40	42.40	35.60
	b'	82.80	76.70	73.80	74.30	74.10	72.40	71.30	70.50	69.80
	b"	-43.00	-43.60	-44.30	-44.60	-47.40	-48.40	-49.10	-59.50	-49.90
Total in-plane slip at rivet-panel interface (mm)	g	159.70	152.50	150.70	150.70	159.30	166.90	176.60	187.50	199.30
	h	66.40	75.30	82.20	84.70	101.40	114.70	-27.50	142.90	157.60
Total out-of-plane slip (mm)	c	-25.20	-24.10	-25.50	-26.20	-33.30	-32.40	-33.42	-36.04	-39.79
	c'	-29.70	-24.78	-21.57	-20.07	-16.30	-18.64	-22.12	-26.52	-31.42
	c"	-8.48	-5.60	-3.31	-1.69	5.44	4.79	2.03	-2.15	-7.21
	d"	2.10	2.50	2.80	3.00	2.20	3.20	4.30	5.80	8.70
Peak tensile stress in elements adjacent to rivet hole (MPa)	e	1422.0	1304.4	1243.1	1254.2	1323.0	1330.6	1362.1	1409.2	1459.9
	e"	634.6	623.7	607.0	610.3	721.0	763.2	864.3	933.2	998.4
	f	1374.6	1267.6	1231.2	1232.4	1263.5	1273.2	1291.1	1314.9	1345.8
	f"	635.7	629.6	640.0	655.8	748.1	818.4	894.2	967.7	1038.5
Peak Contact Pressure (MPa)	Shank-hole (A)									
	Panel-Panel (B)	494.2	0.0	0.0	552.2	754.0	866.6	870.6	787.2	628.8
Contact Pressure (MPa)	Panel-Rivet head (C)									

TABLE 12.10. JOINT EXCESS COMPLIANCE AND RIVET TILT AT SPECIFIED VALUES OF APPLIED STRESS

Joint Type	t (mm)	P_1 (mm)	Clamping	Applied Stress (MPa)	Joint End Displacement (μm)	C' (m/GN)	C'' (m/GN)	C, Excess Compliance (m/GN)	Rivet Tilt (°)	Model No.
Al lap joint	1.53	30.6	0%	30	180	128.6	93.4	35	2.5	4S-29
Al lap joint	1.53	30.6	0%	125	--	--	--	27.5	3.6	4S-1
Al lap joint	1.53	30.6	0.5%	30	166	118.6	93.4	25	2.4	4S-30
				42	231	117.2	93.4	24	--	--
				54	295	116.6	93.4	23	--	--
Al butt joint	3.06	30.6	0%	30	307.2	109.3	93.4	16	0.6	3S-1
Al butt joint	3.06	30.6	0.5%	30	276.3	98.3	93.4	5	0.5	3S-2
				42	394.4	100.4	93.4	7	--	--
				60	575.2	102.4	93.4	9	--	--
				69	665.6	103	93.4	10	--	--
Steel butt joint	3.06	30.6	0.85%	125.8	228.3	19.4	17.5	2	0.1	4S-1
				200	378.2	20.2	17.5	3	--	--
				220	420.0	20.4	17.5	3	--	--
				321	633.4	21.1	17.5	4	--	--

SECTION 3

CALCULATIONAL MODELS AND VALIDATION

FINITE ELEMENT MODELS

This chapter presents the details of the finite element models used for the calculations tabulated in Appendix A. It includes information about the geometries of all the models. Differences among the models (e.g. friction coefficients) are given in the tables in Appendix A; Chapter 15 presents the details of the material behavior used for all of the models.

All of the finite element models used for the data presented fall into the following three categories:

- 2D models of butt joints, attachment joints and open hole panels [18, 19]
- 3D models of single rivet-row lap and butt joints (including standard and countersunk rivet heads, interference and clamping, sealants and adhesives, biaxial loading) (1, 13-16)
- 3D models of double rivet-row lap, butt and doubler joints (standard and countersunk rivet heads) [1, 13, 14, 19, 20, 35]

For each category, a basic model was developed and modified as needed. The basic models will be described below, followed by the details of the various modifications. All models were created and run using the finite element code ABAQUS.

13.1 2D models of butt joints, attachment joints and open hole panels

Two types of butt joint models have been developed: 2D and 3D (described in Chapter 13.2). The 2D model can also be used to represent pinned butt and attachment joints and open hole panels.

The two-dimensional pinned connection finite element model used to represent butt joints is shown in Fig. 13.1. Only the center panel is treated and interaction with the side panels is neglected. For calculations treating a finite plate, the pin was held fixed at its center while a tensile stress of 60 MPa was applied at the far edge of the plate along its longitudinal (2–) direction. This stress corresponds to a load per unit thickness of the plate, $Q = 1836$ kN/m for the 30.6 mm panel width. For an infinite plate, a statically equivalent concentrated load, $Q = 1836$ kN, was applied to the center of the pin while the far edge of the plate was held fixed in the 2-direction. In addition, to represent the case where there is a row of pins, and not a single pin, some calculations were performed on the finite plate with its sides held fixed in the lateral (1–) direction, in addition to the longitudinal, uniaxial stress. Both the pin and plate were considered to be in plane strain. The panel is 30.6 mm wide and 168.3 mm long. The center of the hole (and pin) is 15.3 mm from the lower edge and each of the sides of the plate. The radius of the hole is $R = 3.06$ mm. Only the case of equal pin and

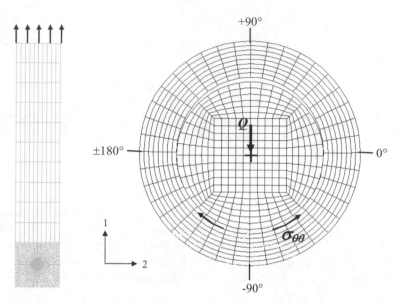

Figure 13.1. Two-dimensional finite element model of the pinned connection. A section of the deformed mesh near the contact region is shown on the right. The angular location convention and tangential stress around the hole, $\sigma_{\theta\theta}$, are also shown, $\sigma_{\text{nom}} = 125$ MPa.

hole radii was considered. The pin/hole radius, R, and finite sheet dimensions were derived from typical airframe riveted connections. A Coulomb friction model was assumed.

Attachment joints have the same geometry as butt joints, but are held in place by one or more rigid fasteners; i.e. the fasteners and support structure are rigid compared to the panel.

Open hole panel models consist of the panel shown in Fig. 13.1 without the pin.

Models Which Include Biaxial Loading. In addition to the pin being fixed at its center, a tensile stress, $\sigma_L = 125$ MPa, is applied at the far edge of the panel in the longitudinal (1–) direction as shown in Fig. 13.1. This is considered to be the principal loading direction. In addition, a transverse (secondary) load, σ_T, is also applied. Three types and values of transverse loading are considered: (i) Case 1, in which the sides of the panel are held fixed in the transverse (2–) direction. This is the symmetry boundary condition imposed by the two adjacent, identical units, (ii) Case 2, in which a tensile stress of 62.5 MPa is applied in the transverse direction along the sides of the panel, and (iii) Case 3, in which a fixed displacement of ±13.66 μm is applied all along the sides of the panel. The longitudinal and transverse loads are applied simultaneously in all cases.

13.2 3D models of single rivet-row lap and butt joints (standard and countersunk rivet heads)

The basic single rivet-row lap joint model is shown in Figure 13.2. The model geometry consists of two partially overlapping sheets (panels) coupled by a rivet. This represents one half

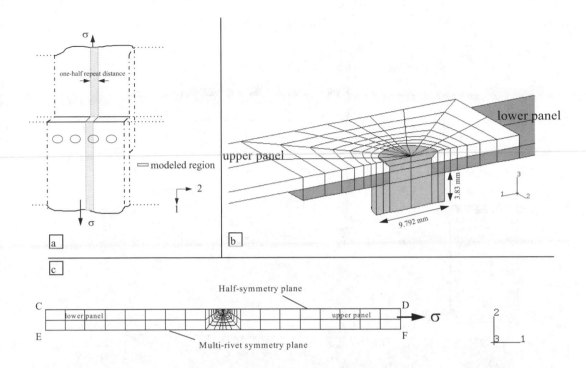

Figure 13.2. (a) Multi-riveted, single-row lap joint, of which one-half unit is considered for the 3-D half-symmetry finite element model (b) the thickness of each panel is t = 1.53 mm; the rivet shank diameter is 6.12 mm; the rivet head diameter and height are 9.792 mm and 3.83 mm respectively (c) plan view of mesh. The overall length of the model is 306 mm; the length of the overlap region is 30.6 mm; the width of the model (half the repeat distance) is 15.3 mm.

unit of a multi-riveted, single row, long panel extending infinitely in the positive and negative 2-directions. The repeat distance between successive units is 30.6 mm. The upper panel may possess a countersunk hole. Finite element analysis (FEA) was performed using the ABAQUS code; 27-noded brick elements were used to mesh the three bodies, and single-noded slide surface elements were defined internally to solve the contact inequality constraints. The four rivet geometries considered are: (1) A double headed rivet with no countersinking, (2) A rivet with one standard head at its lower end and one head countersunk to a depth of half the thickness of a single sheet, the included angle being 100°, (3) A rivet with one standard head at its lower end and one head countersunk to a depth of the thickness of a single sheet. In this case, the maximum and minimum diameter remain unchanged, resulting in an included countersink angle of 61.6°, and, (4) A rivet with one standard head at its lower end and one head countersunk to a depth of the thickness of a single sheet which contains a 100° angle.

A nominal tensile stress is applied at the non-lapping end face (DF) of the upper sheet, while the corresponding face belonging to the lower sheet (CE) is fixed along the 1-axis. Figure 13.2 shows the main features of the models. For the infinite panel, planes (1-3) CD and EF are symmetry planes and are constrained against motion in the 2-direction. In

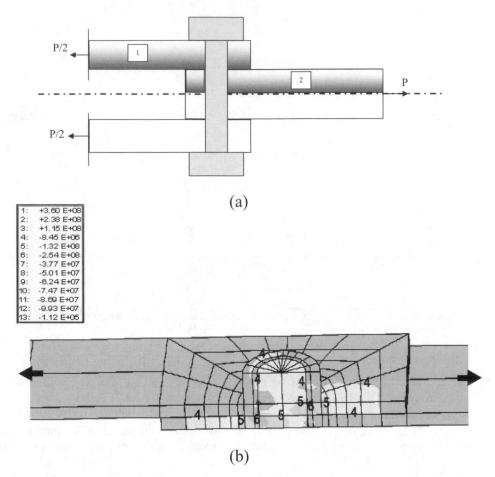

1:	+3.60 E+08
2:	+2.38 E+08
3:	+1.15 E+08
4:	-8.45 E+06
5:	-1.32 E+08
6:	-2.54 E+08
7:	-3.77 E+07
8:	-5.01 E+07
9:	-6.24 E+07
10:	-7.47 E+07
11:	-8.69 E+07
12:	-9.93 E+07
13:	-1.12 E+05

(b)

Figure 13.3. (a) Schematic of the 3D butt joint model as derived from the existing lap joint model including only the shaded portion of the sheets and half the rivet, (b) σ_{11} stress contours for the 3D butt joint model (deformations are magnified 3x, applied stress is 60 MPa).

addition, the edges parallel to the 2-axis, and on 2-3 planes DF and CE are fixed in the 3-direction to prevent unrestrained rotation of the body in space. All the boundary conditions are symmetric. For finite-width panels, the multi-rivet symmetry plane (1-3) along EF, shown in Figure 13.2, is further constrained against motion in the 1 and 3-directions, so that it is no longer a symmetry plane.

Values of the friction coefficients used in the various models along with applied nominal tensile stress are presented in the tables in Appendix A.

The finite element meshes might typically consist of 3171 user defined nodes, 252 user defined elements and 609 internally generated contact elements. A typical figure for the total number of variables is 11,340.

Models Which Represent Butt Joints. Figure 13.3a shows a schematic of the butt joint model. The shaded portion and half the river are considered in the model. This model was

**TABLE 13.1. THE FINITE ELEMENT MODELS WHICH INCLUDE
INTERFERENCE AND CLAMPING**

Model	Rivet Heads	% Shank Radial Interference	% Shank Height Interference (Clamping)
4S-19	2 standard, non-countersunk	0	0
4S-20	2 standard, non-countersunk	1	0
4S-21	2 standard, non-countersunk	2	0
4S-22	2 standard, non-countersunk	0	0.5
4S-23	2 standard, non-countersunk	1	0.5
4C-24	1 standard 1 countersunk (100^0) to half panel thickness	0	0
4C-25	1 standard 1 countersunk (100^0) to half panel thickness	1	0
4C-26	1 standard 1 countersunk (100^0) to half panel thickness	2	0
4C-27	1 standard 1 countersunk (100^0) to half panel thickness	0	0.5
4C-28	1 standard 1 countersunk (100^0) to half panel thickness	2	0.5

derived from the existing lap joint model by deleting the lower head of the rivet. The loads are shown for the resulting butt joint. Since the cross-sectional area of the central sheet is trwice that of the abutting sheets, the nominal strress in all the sheets is identical. For 0.5% clamping, a nominal applied stress of 60 MPa and frictionless contsct, the net section peak stess is 330.8 Mpa (SCF = 5.5) and the fross section peak stress ie 67.4 MPa (SCF = 1). Both these values are obtained from sheet #1. The peak net section stress is in sheet #1; the peak gross section stress could be in sheet #2. The length of sheet #1 can be easily shortencd for future calculations. Figure 13.3b shows the σ_{11} stress contours for this analysis.

Models Which Include Interference And Clamping. Table 13.1 indicates the values of initial residual stresses considered for the interference and clamping analyses. Forcing radial conformity between an initially oversized rivet shank and the panel holes simulated the effects of rivet-panel interference stresses. In this case the interference is produced only in the radial direction. Clamping stresses were generated by forcing the elongation of a rivet whose initial shank height is slightly less than the combined depth of the two panels ($z = 2t$). The clamping forces were also introduced by an interference (misfit) method. Table 13.1 indicates the individual amounts of interference and clamping misfits analyzed, where the misfit is expressed as a percentage of the panel hole diameter (6.12 mm) or combined panel thickness (3.06 mm).

For these analyses a nominal, remote cyclic load defined by $\Delta\sigma = 90$ MPa and $R = 0.1$ was applied at the non-lapping end face of the upper sheet while the corresponding face

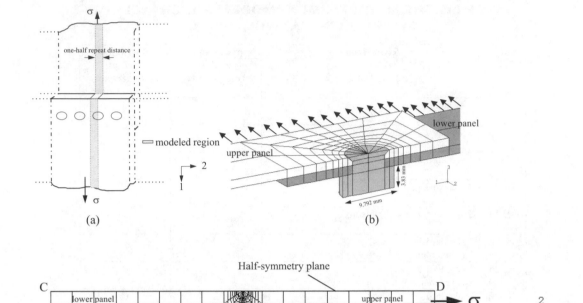

Figure 13.4. (a) Multi-riveted, single-row lap joint, of which one-half unit is considered for the 3-D half-symmetry finite element model (b) the thickness of each panel is $t = 1.53$ mm; the rivet shank diameter is 6.12 mm; the rivet head diameter and height are 9.792 mm and 3.83 mm respectively (c) plan view of mesh. The overall length of the model is 306 mm; the length of the overlap region is 30.6 mm; the width of the model (half the repeat distance) is 15.3 mm.

belonging to the lower sheet was fixed along the x-axis; the Cartesian systems (1,2,3) and (x,y,z) are equivalent.

Ball-on-flat fretting tests conducted for aluminum rubbing against itself have shown the friction coefficient, μ, to vary from 0.2 to 0.5 [20]. The slip amplitude in these tests ranged between 10 μm and 80 μm and the normal load between 28.2 N and 34 N; the frequency was 1 Hz. For this study, the variation in μ was neglected, and the Coulomb friction model with a constant value of $\mu = 0.4$ was assumed for all interfaces.

Models Which Include Sealants and Adhesives. The modeling of sealants and adhesives was performed using the TALA method; this approach is described in detail in Chapter 14. These studies used the basic single rivet-row lap joint model described above with some variations in nominal applied load, panel thickness and material properties. These details are presented in Table A.9.

Models Which Include Biaxial Loading. Figure 13.4 shows the additional loading applied along the panel length for the biaxial case. Table 13.2 indicates the magnitudes of transverse

TABLE 13.2. THE FINITE ELEMENT MODELS

Model	Transverse Boundary Condition	Rivet Heads
4S-16	Both panels' sides constrained	2 standard
4S-17	Transverse stress applied to panel sides	2 standard
4S-18	Transverse displacement applied to panel sides	2 standard
4C-17	Both panels' sides constrained	1 standard, 1 countersunk (100°) to half panel thickness
4C-18	Transverse stress applied to panel sides	1 standard, 1 countersunk (100°) to half panel thickness
4C-19	Transverse displacement applied to panel sides	1 standard, 1 countersunk (100°) to half panel thickness

loading; three types and values of transverse loading are considered: (i) Case 1, in which the sides of the panel are held fixed in the transverse (2–) direction. This is the symmetry boundary condition imposed by the two adjacent, identical units, (ii) Case 2, in which a tensile stress of 62.5 MPa is applied in the transverse direction along the sides of the panel, and (iii) Case 3, in which a fixed displacement of ±13.66 μm is applied all along the sides of the panel. The longitudinal and transverse loads are applied simultaneously in all cases.

13.3 Double rivet-row lap joints (standard and countersunk rivet heads)

The model geometry consists of two partially overlapping sheets (panels) fastened by two two-headed, non-countersunk rivets, shown in Fig. 13.5. This represents one half unit of a multi-riveted, double row, long panel extending infinitely in the positive and negative 2-directions. The repeat distance between successive units is 30.6 mm. Finite element analysis (FEA) was performed using the ABAQUS code; twenty-seven noded brick elements were used to mesh the three bodies and single-noded slide surface elements were defined internally to solve the contact inequality constraints. Two rivet geometries are treated: (1) A double headed rivet with no countersinking, and, (2) A rivet with one standard head at its lower end and one head countersunk to a depth of the thickness of a single sheet which contains a 100° angle.

A nominal tensile stress is applied at the non-lapping end face of the upper sheet, while the corresponding face belonging to the lower sheet is fixed along the 1-axis. Figure 13.5 shows the main features of the models. Planes (1-3) CD and EF are symmetry planes and are constrained against motion in the 2-direction. In addition, the edges parallel to the 2-axis on (2-3) planes DF and CE are fixed in the 3-direction to prevent unrestrained rotation of the body in space. All the boundary conditions are symmetric for infinitely wide panels.

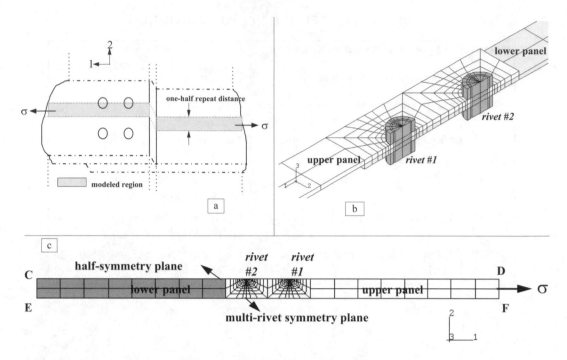

Figure 13.5. (a) Multi-riveted, double-row riveted lap joint, of which one-half unit is considered for the 3-D half-symmetry finite element model (b) the model consists of two rivets (#1 and #2) installed in two holes (#1 and #2 respectively). The thickness of each panel is t = 1.53 mm. The non-countersunk rivet shank diameter is 6.12 mm and the countersunk rivet diameter is 7.94 mm; the rivet head diameter and height are 9.792 mm and 3.83 mm respectively (c) plan view of mesh. The overall length of the model is 336.6 mm and the width of the model is 15.3 mm (half the repeat distance); the length of the overlap is 61.2 mm.

Values of the friction coefficients used in the various models along with applied nominal tensile stress are presented in the tables in Appendix A.

A typical finite element mesh for a double rivet-row lap joint consists of 5832 user defined nodes, 472 user defined elements and 1213 internally generated contact elements. The total number of variables is 21,135.

THIN ADHESIVE LAYER ANALYSIS (TALA) FOR MODELING SEALANTS AND ADHESIVES INSTALLED IN JOINTS

14.1 DEVELOPMENT OF A THIN ADHESIVE LAYER ANALYSIS (TALA)

Modeling of a sealant layer using FEA is not practical because the thickness of the sealant layer is so small compared to the panel dimensions. TALA avoids this problem by representing the sealant using spring elements connecting two contacting surfaces, eliminating solid element representation of the sealant layer in the finite element models (see Figure 14.1). Thus, each pair of coincident nodes between two contacting surfaces in the finite element models is connected by spring elements (defined as SPRING2 element in ABAQUS). The mechanical behavior of the springs must be defined in local normal and local shear directions in the finite element models. The stresses and strains in the springs are converted to forces and displacements interacting between the two contacting surfaces. The properties of the adhesive are treated as isotropic and time and load rate independent. This chapter is based on References [6] and [21]. The nomenclature for this chapter is listed at the end of the chapter.

14.2 CONVERSION OF ADHESIVE SOLID ELEMENT TO SPRING ELEMENT FOR 2-D FEM

Two methods are used for changing the solid adhesive element to spring elements: one for interfacing 4-noded elements and the other for 8-noded elements. Each method requires different values for A_i as described below.

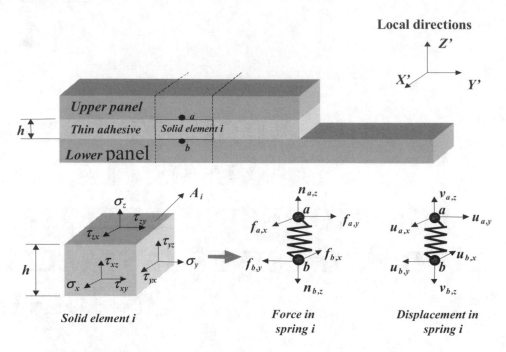

Figure 14.1. Schematic of spring representation of solid adhesive element.

1. **One solid element to one spring method** (for 4-noded elements). This method is used to change one solid adhesive element to one spring as shown in Figure 14.2(a). It changes adhesive element i (bonding elements U_i and L_i) to spring S_i. A spring S_i is composed of two spring elements, connecting node u_i with node l_i, a normal and a shear spring element. The area for spring S_i is equal to

$$A_i = w_i \cdot d \tag{1}$$

Where　　w = the width of each element (in y'-direction) along the interface line
　　　　　d = the depth of each element (in x'-direction); this value is always one unit for two-dimensional plane strain.

2. **One solid element to three spring method** (for 8-noded elements). This method is used to change one solid adhesive element to three springs, as shown in Figure 14.2(b). It changes adhesive element i (bonding panel elements U_i and L_i) to springs S_{i-1}, S_i and S_{i+1}. The spring S_{i-1} is composed of two spring elements connecting node u_{i-1} with node l_{i-1}, a normal and a shear spring element. The area for spring S_{i-1} is:

$$A_i = w_i \cdot d \tag{2}$$

Where w = is the width of each element (in y'-direction) and d = depth of each element (in x'-direction). Similarly, the spring S_i is composed of two spring elements connecting node u_i with node l_i in the normal and shear directions. The area for spring S_i is

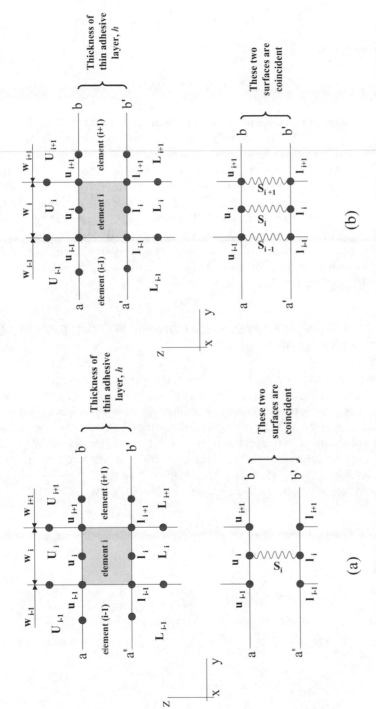

Figure 14.2. Schematic of springs between two interfacing surfaces for two dimensional finite element model (a) using one spring, and (b) using three springs.

199

$$A_i = (w_i/2) \cdot d \tag{3}$$

The area for spring S_{i+1} is

$$A_{i+1} = [(w_{i+1}/4) + (w_i/4)] \cdot d \tag{4}$$

Conversion of adhesive solid element to spring element for 3-D FEM. For the three-dimensional finite element model, the change from thin solid adhesive elements to spring elements is based on the same idea used for the two-dimensional problem; however, the area for each spring (A_i) must be calculated differently. Figure 14.3 shows the features of the three-dimensional finite element with elements U_i and L_i bonded by a thin adhesive layer. Nodes on surface a-b-c-d of element U_i and a'-b'-c'-d' of element L_i are coincident. The thin adhesive element i bonding the two surfaces of elements U_i and L_i are changed to 9 springs for use with 27 node isoparametric elements, etc. Spring (S_i) connects nodes u_i and l_i together, etc. Each spring is composed of three spring elements, one normal spring, and two shear springs. Areas D_i through D_{i+8} are shown schematically in Figure 14.3 and are used to obtain A_i through A_{i+8} for the springs, as shown in Table 14.1.

14.3 CONVERSION OF STRESS-STRAIN RELATIONSHIP TO FORCE-DISPLACEMENT RELATIONSHIP FOR THIN ADHESIVE LAYER

The adhesive layer is assumed to be very thin. Thus, the stresses σ_x, σ_y, τ_{xz}, τ_{xy}, τ_{yz}, τ_{yx}, γ_{xz}, γ_{xy}, γ_{yz}, and γ_{yx}, as shown in Figure 14.1 are neglected using TALA. Only σ_z, τ_{zx} and τ_{zy} act between the two nodes. The normal stress (σ_z) is changed to normal force, while the shear stresses (τ_{zx}, τ_{zy}) are changed to shear force in each spring element. The normal and shear strains (ε_z, γ_{zy}, γ_{zx}) are changed to normal and shear relative displacements in spring i.

The equations used for converting the stresses and strains in the solid element to forces and displacements to define the properties of the spring element are presented below. In the normal direction (tension and compression spring), by Hooke's law:

$$\sigma_z = E^*(\varepsilon_z) \tag{5}$$

$$(n_{a,z} + n_{b,z})/A_i = E^*(\varepsilon_z) \tag{6}$$

$$\varepsilon_z = (v_{a,z} + v_{b,z})/h \tag{7}$$

$$(n_{a,z} + n_{b,z}) = (E^*(A_i/h))^* (v_{a,z} + v_{b,z}) \tag{8}$$

$$(n_{a,z} + n_{b,z}) = K_n^*(v_{a,z} + v_{b,z}) \tag{9}$$

$$F_{n,i} = K_{n,i} * v_{n,i} \tag{10}$$

$$K_{n,i} = E^*(A_i/h) \tag{11}$$

Where $F_{n,i}$ is the normal force transmitted in spring element i, $v_{n,i}$ is the relative displacement of spring element i in the normal direction, $K_{n,i}$ is the local stiffness of spring element i in the

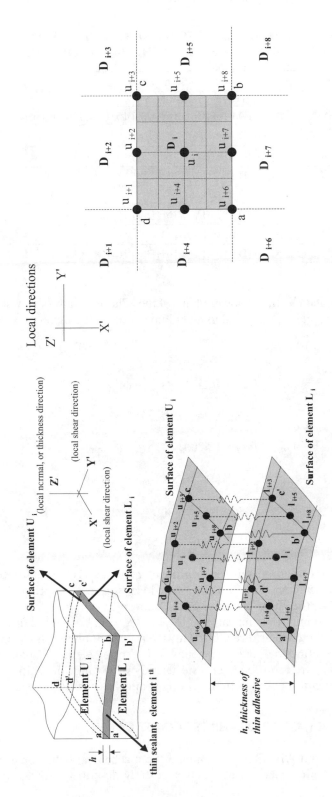

Figure 14.3. Schematic for representing solid adhesive element with spring elements.

201

TABLE 14.1. FORMULATION FOR CALCULATING THE SPRING AREA FOR THE 3-D MODEL

Spring	Spring Area
S_i	$A_i = (D_i)/16$
S_{i+1}	$A_{i+1} = (D_i)/16+(D_{i+1})/16+(D_{i+2})/16+(D_{i+4})/16$
S_{i+2}	$A_{i+2} = (D_i)/8+(D_{i+2})/8$
S_{i+3}	$A_{i+3} = (D_i)/16+(D_{i+2})/16+(D_{i+3})/16+(D_{i+5})/16$
S_{i+4}	$A_{i+4} = (D_i)/8+(D_{i+4})/8$
S_{i+5}	$A_{i+5} = (D_i)/8+(D_{i+5})/8$
S_{i+6}	$A_{i+6} = (D_i)/16+(D_{i+4})/16+(D_{i+6})/16+(D_{i+7})/16$
S_{i+7}	$A_{i+7} = (D_i)/8+(D_{i+7})/8$
S_{i+8}	$A_{i+8} = (D_i)/16+(D_{i+5})/16+(D_{i+7})/16+(D_{i+8})/16$

local normal direction, and E is the elastic modulus of the adhesive. In the shear direction (shear spring) in x'-y' plane τ_{zx} and τ_{zy} are equal in magnitude because the material is isotropic.

$$\tau_{zx} = G^*(\gamma_{zx}) \tag{12}$$

$$(f_{a,x} + f_{b,x})/A_i = G^*(\gamma_{zx}) \tag{13}$$

$$\gamma_{zx} = (u_{a,x} + u_{b,x})/h \tag{14}$$

$$(f_{a,x} + f_{b,x}) = (G^*(A_i/h))^*(u_{a,x} + u_{b,x}) \tag{15}$$

$$(f_{a,x} + f_{b,x}) = K_f^*(u_{a,x} + u_{b,x}) \tag{16}$$

$$F_{f,i} = K_{f,i}^* u_{f,i} \tag{17}$$

$$K_{f,i} = G^*(A_i/h) \tag{18}$$

Where $F_{f,i}$ is the shear force transmitted in spring element i, $u_{f,i}$ is the relative displacement of spring element i in the shear direction, $K_{f,i}$ is the local stiffness of spring element i in the shear direction, and G is the shear modulus of the adhesive.

For the case of a linear material, the values of $K_{n,i}$ and $K_{f,i}$ can be defined directly for spring element i in the normal and shear directions. However, for a nonlinear material $K_{n,i}$ and $K_{f,i}$ vary. Then, in order to define the nonlinear behavior for spring element i, pairs of force-relative displacement values are required over a sufficiently wide range of relative displacement values. For more information about defining spring properties see Reference [86].

14.4 MECHANICAL BEHAVIOR OF THIN SEALANT LAYER

In order to use TALA and ABAQUS to simulate the mechanical behavior of the layer, the actual behavior of the thin sealant coating must be fully described. Low weight aerospace

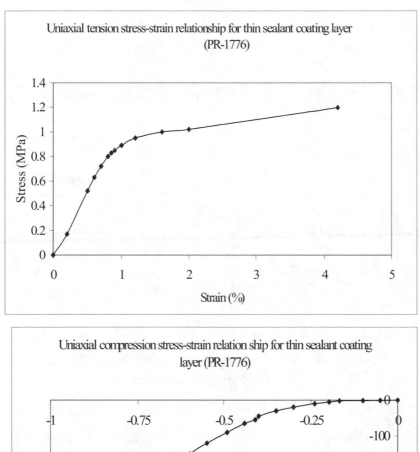

Figure 14.4. The uniaxial stress-strain relationship for a thin sealant coating layer 170 μm thick.

sealant (PR-1776 B-2 model 654) supplied by PRC-DeSoto International, Inc. is used here. In the normal direction, the stress-strain relationship for a thin sealant coating layer of this material subjected to uniaxial static tension and compression loading is obtained by Fongsamootr [7], as shown in Figure 14.4. In the shear direction, data for the sealant, subjected to combined torsion and compression loading, are obtained by Kamnerdtong [8] as shown in Figure 14.5.

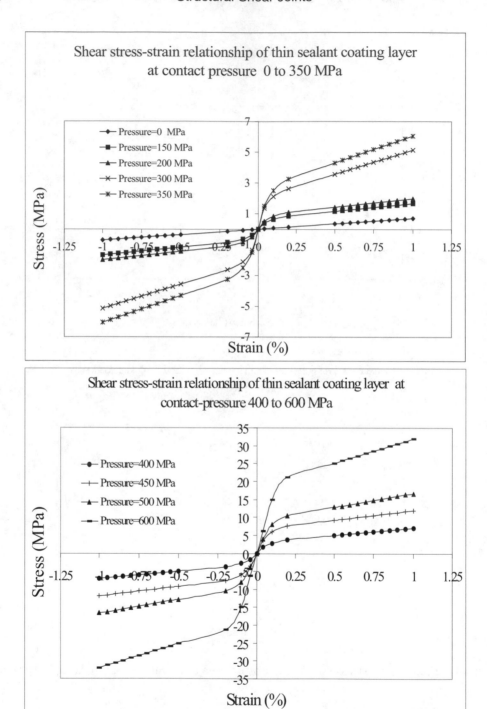

Figure 14.5. The shear stress-strain relationships for a thin sealant coating layer 100 μm thick.

14.5 SHEAR SPRING DEPENDENCE OF CONTACT PRESSURE

A procedure for the determination of the pressure dependent stress-strain behavior for the shear spring is described here. Figure 14.5 gives shear stress-strain relationships for various pressures (in-between values are obtained through interpolation). To start with, the contact pressure is assumed to be zero. After the first ABAQUS run, the normal force in each spring is output (DAT file). If the value of the normal force ($F_{n,i}$) is tensile or zero, the force-displacement relationship for the shear spring is kept constant. This is because shear behavior data was not obtained as a function of tension, thus the assumption is made that tension has a negligible affect on the sealant shear properties. If the value of the normal force ($F_{n,i}$) is compressive, the compressive force is divided by the small area (A_i) to get the new value of the contact pressure (P_i) for that spring element. Then, for the next run, the shear spring of that element is redefined to have a new shear force-displacement relationship from the new contact pressure. Using this idea, the model is run repeatedly and the shear spring is redefined until the new contact pressure and previous contact pressure for each spring converge. EXCEL is used to provide a new input file for each new run and to define new shear spring pressure dependence following P_{inew} for each spring.

Symbols

A_i = Nominal area for i^{th} spring

E = elastic modulus of adhesive

$F_{n,i}, F_{f,i}$ = normal, shear force transmitted in i^{th} spring element

G = shear modulus of adhesive

$K_{n,i}, K_{f,i}$ = local stiffness of ith spring element in the local normal, shear directions

n, f = normal, shear forces

P_i = contact pressure for i^{th} spring

S_i = i^{th} spring

$u_{f,i}$ = relative displacement of ith spring element in shear direction

u_i, l_i = nodes connected by i^{th} spring

U_i, L_i = Elements bonded by i^{th} spring

$v_{n,i}$ = relative displacement of ith spring element in the normal direction

w_i = width of ith element (in y'-direction) along the interface line

σ, γ = normal, shear stresses

MATERIAL MODELS FOR THE FINITE ELEMENT CALCULATIONS

This chapter presents the details of the material behaviors used in the finite element and TALA calculations. The sealant properties for the TALA calculations are described in Figures 14.4 and 14.5.

15.1 Materials used in the finite element calculations

There are a total of 5 pin-panel material combinations used in the calculations summarized in Chapter 7:

(1) Aluminum pin-Aluminum panel
(2) Steel pin-Aluminum panel
(3) Titanium pin-Aluminum panel
(4) Aluminum pin-Epoxy panel
(5) Steel pin-Steel panel

Of these, the only non-linear treatments are for the Aluminum-Aluminum models for which most are linear with a few examples of elastic-isotropic-plastic (EIP) and elastic-linear-kinematic hardening-plastic (ELKP). The material properties used for all of the uniaxial tension cases, except as noted in Table A.7, are presented in Table 15-1.

In the case of biaxial loading for butt joints, the combinations of linear material properties used are shown in Table 15-2.

TABLE 15.1. SUMMARY OF MATERIAL PROPERTIES FOR UNIAXIAL LOADING CASES

Material	Elastic modulus, E (GPa)	Poisson's ratio, ν	Yield strength (MPa)	Plastic modulus, M (slope, GPa)
Aluminum (linear)	70	0.30	--	--
Aluminum (EIP) AA7075-T6	70	0.30	531	0.70
Aluminum (ELKP)	70	0.30	369.9	57.5
Steel	207	0.25	--	--
Titanium	117	0.31	--	--
Epoxy	3	0.36	--	--

TABLE 15.2. SUMMARY OF MATERIAL PROPERTIES IN BIAXIAL LOADING CASES FOR BUTT JOINTS

Material	E_1 (GPa)	E_2 (GPa)	ν_1	ν_2
Al-Al	70	70	0.31	0.31
St-Al	207	70	0.33	0.31
Ti-Al	117	70	0.31	0.31
Al-Epoxy	70	3	0.31	0.36

VALIDATION OF FINITE ELEMENT CALCULATIONS AND TALA

16.1 FINITE ELEMENT MODEL VALIDATION

The validity of the finite element analyses has been tested in a number of ways:

1. The 2D finite element model of a large panel with an empty, open hole was compared with the closed form solution described by Timoshenko [29]. The SCF derived from the finite element model, SCF = 3.0 is identical to the closed form solution value.

2. The 2D finite element model of a pinned connection was compared with strain gage and photoelastic measurements [87]. Table 16.1 summarizes the calculated and measured SCF-values. The two sets of results are in good accord when the differences in P_1/D are accounted for.

3. The 2D finite element analysis of a frictionless, pinned connection [1,18] was compared with the corresponding, closed form solution of Cavaella and Decuzzi [25]. As illustrated in Figure 16.1, the finite element model accurately reproduces the pressure distribution of the closed form solution.

4. To test the adequacy of the 3D model mesh refinement, the mesh was employed to analyze a finite, pinned connection with friction under in plane loading. The results are compared in Figure 16.2 with those obtained with the more refined mesh of the 2D finite element analysis. The two sets of results for contact pressure and slip distributions are in good agreement. The difference in SCF obtained using the 2D and less refined 3D mesh under in-plane loading is about 7%.

5. To test the adequacy of the 3D finite element analysis for treating out-of-plane distortions an analysis of a single, rivet-row lap joint was compared with experimental measurements of the joint in-plane and out-of plane displacements [1,16]. The results are summarized in Figure 16.3. The calculations predict the variations of displacement with load and the absolute values of displacement within 10%.

6. A further test of the 3D finite element analysis of a single, rivet-row lap joint is obtained by comparing the results of Fung and Smart [88,89] with another 3D finite element model of a different lap joint. Table 16.2 illustrates that the finite element calculations produce comparable SCF values when differences in P_1/D and D/t are taken into account.

TABLE 16.1. COMPARISON OF THE MEASURED STRESS ELEVATION IN A BUTT JOINT [87] WITH FINITE ELEMENT ANALYSIS VALUES [18]

METHOD	PANEL/FASTENER	P_1/D	$\theta\,°$	σ/σ_1 or SCF
Strain gages	Steel/Steel	3.76	0°	5.1
		5.0[a]	0°	5.5
Photoelastic	Fosterite/Fosterite	4.2	0°	5.0
		4.2	-14°	6.3
Finite element analysis (Model 3P-16)	Aluminum/Aluminum	5.0	4°	6.6
		4.2[a]	4°	6.3

[a] Estimate for $P_1/D=5$ obtained with Figure 5.2.

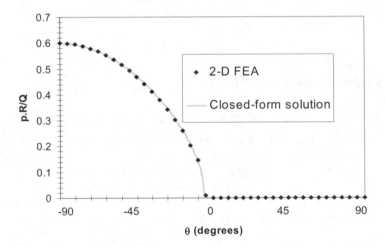

Figure 16.1. Finite element and analytical solutions for the contact pressure distribution in a pinned connection with an infinite sheet.

16.2 VALIDATION OF THE THIN ADHESIVE LAYER ANALYSIS (TALA)

The validity of TALA was tested by comparing its predictions with two previous analytical studies and with experimental measurements.

Analysis of Reinforced Cantilever Beam. Cornell [90] has analyzed the response of a filleted brazed tab specimen: two cantilevered steel beams bonded using a brazed layer (see Fig. 16.4). Cornell's model consisted of two linear elastic beams bonded with an infinite number of linear tension and shear springs. In order to verify TALA, the brazed tab specimen was analyzed as a two-dimensional plane strain finite element model with no fillet. The model uses 150 elements for the steel tab and 570 elements for the steel base bar. The braze bonding layer was transformed into 51 linear tension-compression spring elements and 51

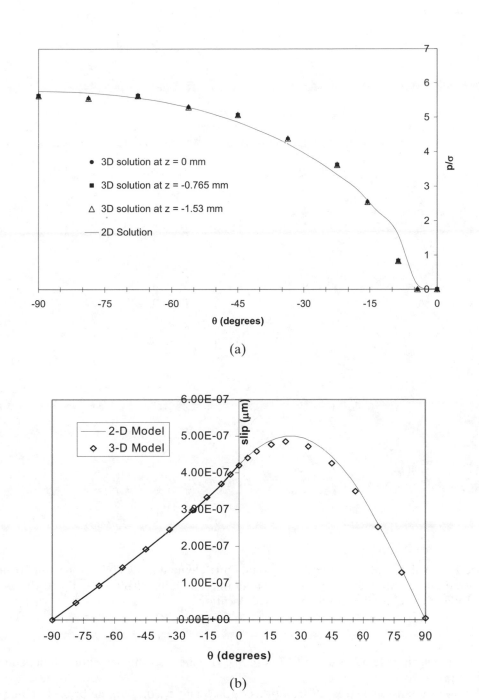

(a)

(b)

Figure 16.2. Comparisons of computed distributions of (a) the normalized contact pressure, p/σ and (b) interfacial slip distance obtained from the two-dimensional (2D) and three-dimensional (3D) finite element models.

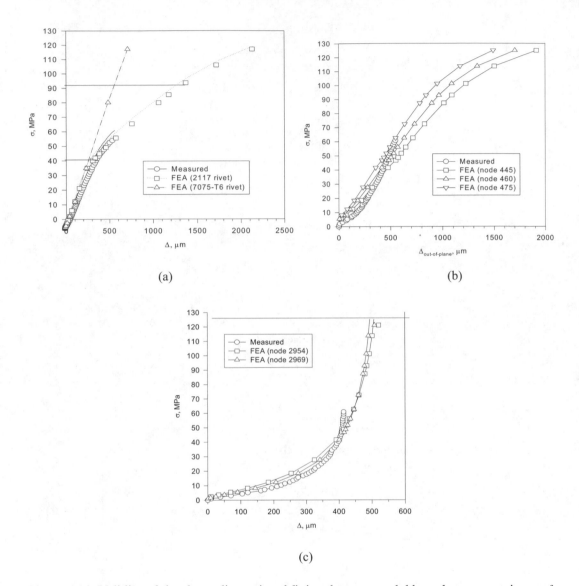

Figure 16.3. Validity of the three-dimensional finite element model based on comparisons of measured and computed values of (a) overall in-plane joint displacement, and (b) and (c) out-of-plane panel movement near the rivet location.

linear shear spring elements using TALA [6,21]. The shear spring behavior is independent of contact pressure.

The non-dimensional stress distributions σ_y/σ_{max} and τ_{xy}/σ_{max} in the braze bonding layer are compared in Fig. 16.4 with the analytical closed-form solution obtained by Cornell [90]. It may be seen that the Cornell and TALA analyses are essentially in agreement. Small differences in the magnitude and location of the stresses are due to the fact that TALA does not include the braze layer fillet at the tab ending.

**TABLE 16.2. COMPARISON OF SCFS OBTAINED BY IYER [1, 16]
AND FUNG AND SMART [88]**

Joint F.E. Model	Iyer (1,16)	Fung and Smart (76)
P_1/D	5	~4.4
D/t	4	1.99
$\dfrac{\sigma}{E}\left(\dfrac{L}{t}\right)^2$	~10	~2
SCF Standard head	$6.1 / 8.1^1$	7.3
SCF Countersunk head	$9.8 / 10.7^1$	10.5

[1] Iyer FE model results are adjusted for Fung and Smart joint model geometry using Figures 2.5 and 2.6.

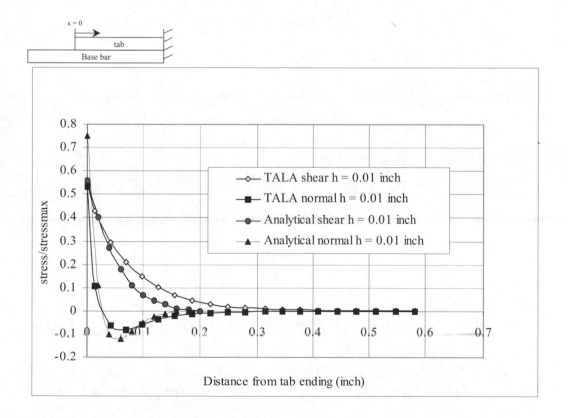

Figure 16.4. Normal and shear stress in braze bonding layer, h (thickness of braze layer) = 0.01.

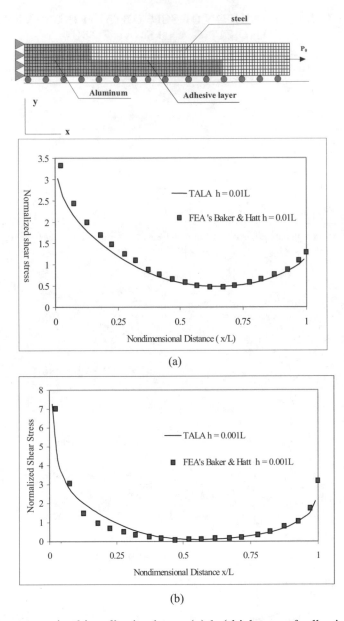

Figure 16.5. Shear stress in thin adhesive layer, (a) h (thickness of adhesive layer) = 0.01L inch (0.254 mm), h = 0.001L inch (0.0254 mm). L is the overlapping length and $0 \leq x \leq L$.

Analysis of Bimetallic Stepped Lap Joint. Baker and Hatt [91] use a linear elastic finite-element analysis to evaluate the behavior of a two-dimensional stepped lap joint (see Fig. 16.5). The aluminum and steel panels are fastened with a thin, linear high modulus adhesive layer. A special element is used which acts like a combined normal and shear spring between two plates (this method differs from TALA in the definition of the connections between coincident and adjacent nodes).

TALA was applied to a two-dimensional plane strain finite element model of the stepped joint using 500 elements for the aluminum panel and 500 for the steel [6,21]. Within TALA, the adhesive layer is transformed to 50 linear tension-compression and 50 linear shear spring elements connecting the lower surface of steel to the upper surface of aluminum using the one spring method, as shown in Figure 14.2(a).

The results of Baker & Hatt are compared with the TALA analyses in Figure 16.5 showing normalized shear stress (τ/p_0) along the overlapping length (x/L) for two different adhesive thicknesses. The TALA analysis results are in agreement with Baker and Hatt's analysis in both shape and magnitude along the overlapping length.

Measurements of lap joints. Experimental validation was obtained by comparing TALA calculations with the measured stress-strain behaviors of 3 aluminum lap joints fastened with a polymer adhesive:

  (i) An ordinary lap joint
 (ii) A stepped lap joint, and
(iii) A single rivet-row lap joint augmented with adhesive.

The joints were fastened with thin, 60–180 μm thick, layers of a low modulus, pressure sensitive, adhesive sealant (PRC-Desoto PR-1776 B-2 model 654). The joint configurations, finite element models and the properties of the aluminum panels and the adhesive are described in detail elsewhere [6,21].

Figure 16.6 shows a comparison of the experimental and finite element relative in-plane and out-of-plane displacement results at the same tensile loads. The calculations agree with the measurements for small applied loads. Small divergences for high loads are probably caused by stress relaxation in the sealant layer during the period of time that the displacements were held constant to allow for reading of the dial gauges.

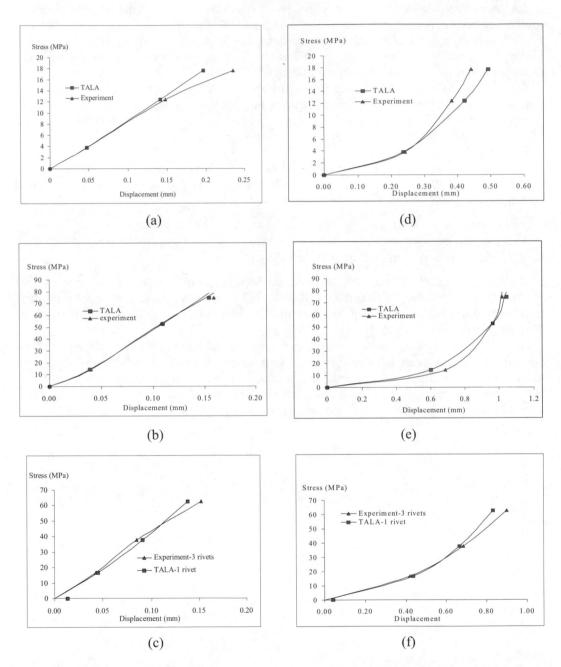

Figure 16.6. (a), (b) and (c) compare the measured and calculated in-plane displacement-nominal stress relation for adhesive lap joint, adhesive stepped lap joint, riveted lap joint, respectively. (d), (e) and (f) compare the measured and calculated out-of-plane displacement-nominal stress relation for adhesive lap joint, adhesive stepped lap joint, riveted lap joint, respectively.

STRESS CONCENTRATION FACTORS IN MULTIPLE ROW JOINTS

17.1 ANALYSES OF MULTIPLE ROW JOINTS

The distribution of fastener loads acting at the several rows of multiple row attachment, butt and lap joints and doublers can be used to estimate SCFs in these joints. Chapter 17.1 outlines a general analytical approach. Chapter 17.2 describes a semi-analytical procedure and 1-dimesional, 2-springs models for estimating the fastener load distribution of joints with 3-dimensional deformation. Chapter 17.3 describes a superposition method which, together with the fastener load distribution and the numerical result for the SCF of single-row joints, provides estimates of the SCFs in multiple row joints. The reliability of the SCF predictions is examined in Chapter 17.4.

17.2 ESTIMATING FASTENER LOADS IN MULTIPLE ROW JOINTS

Equilibrium. The load applied to the panel of a multiple row shear joint is progressively transferred to the facing panel(s) by the successive rows of fasteners. The quantities Q_1, Q_2, and Q_n are the loads transmitted to the fasteners in the first, second and subsequent rows:

$$Q = Q_1 + Q_2 .... + Q_n \qquad (17.1)$$

where $Q - \sigma t P_1$ is the total fastener load. Equation (17.1) is the one-dimensional equilibrium statement. It is convenient to describe these quantities in terms of f_i, the fraction of the total fastener load supported by fasteners in the different rows, i.e., $f_i = Q_i /Q$.

Compatibility. The total extension of the segment of panel between any two specific fastener rows—the distortion of the fastener-fastener hole combination plus that of the panel—must equal the total extension of the facing segment between the same two rows. The fraction, f_i, is then defined by imposing two conditions: (1) the matching of the extensions of

217

facing segments of panel, and (2) static equilibrium. Estimates of the fastener load distribution can be obtained for butt joints with the equilibrium and compatibility conditions described above, This procedure is described by Kulak, Fisher and Struik [51].

In joints which exhibit non-negligible out-of-plane movements, 3D equilibrium and compatibility conditions must be satisfied and a closed-form analysis of the problem is complex. In the next section, a semi-empirical procedure for estimates of the values of f_i for the general case involving three-dimensional panel and fastener deformations is provided.

17.3 ESTIMATING FASTENER LOADS WITH 3D DEFORMATIONS

Under conditions involving three-dimensional deformations, the elastic behavior of the fastener/fastener hole combination can be expressed in terms of the excess compliance, C: the difference in the compliance of a single fastener-row structural joint and the compliance of a single uniform panel with the same overall dimensions (see the introduction to Chapter 12).

Two-Springs Models of Shear Joints. Swift [24], Muller [34] and the present authors have implemented equilibrium and compatibility using the excess compliance idea and the one-dimensional, 2 (linear)-spring models of shear joints shown in Fig. 17.1. One set of springs simulates the fastener-fastener hole response and the other the intervening panel. The models are essentially the same though the terminology used is slightly different. Swift and Muller formulate the problem in terms of C, the compliance of the spring simulating the fastener/fastener hole and C_P, the compliance of the intervening strip of panel. They refer to the two compliance values as the *fastener (or rivet) flexibility* and the *sheet flexibility*. Similar to concepts and models have previously been proposed by Huth [92] and Swift [24]. The present authors have formulated the problem in terms of k and K, the stiffness of the springs simulating the fastener/fastener hole and the intervening strip of panel, and the k/K-ratio, where:

$$k = C^{-1}, K = C_P^{-1}, \text{ and } k/K = C_P/C \ldots \ldots \quad (17.2A, B, and C)$$

These parameters can be evaluated in a number of ways. The quantity C is identical to the excess compliance of a single, rivet-row joint which the authors have evaluated with the finite element models (see Chapter 12 for definition and tabulated values). Alternatively, Huth [92] and Muller [34] have obtained values of C from load displacement measurements of joints. The value, k = 17 MN/m derived from a finite element analysis for a lap joint with countersunk rivets is in accord with the value, k = 14MN/m derived by Huth [92] from measurements for a similar joint. Empirical expressions for C have been formulated by Huth [90,92], Tate and Rosenfield [93] and workers at Boeing [34,94]. The value of K is estimated as:

$$K = \eta E t P_1/P_2 \ldots \ldots \quad (17.3)$$

where E is the elastic modulus of the panel material; t is the panel thickness; P_1 is the fastener pitch (spacing); P_2 is the fastener row spacing, and η accounts for the effects of the none uniform stress distribution and other features neglected by the simplified model. Muller [34] employs the approximation, $\eta = 1$. The authors evaluated the quantity, η, by

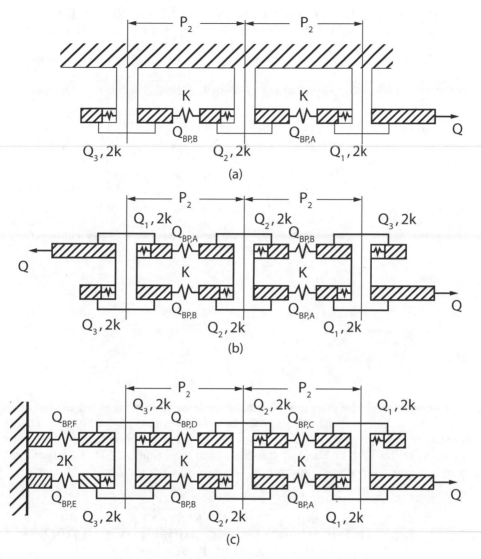

Figure 17.1. Examples of 2-spring models for estimating fastener loads of multiple row, shear joints: (a) Model of a 3-row attachment joint, (b) Model of a 3-row butt or lap joint, and (c) Model of a 3-row doubler.

comparing the fastener load predictions of the 2-spring models of attachment joints with results of finite element calculations. Values of C, k and the k/K-ratio for different types of joints are summarized in Table 17.1 for cases where the panels composing the joint are the same thickness.

Once the values of k and K are known, the fastener loads, f_i, can be estimated solving the appropriate linear 2-spring problem. Huth [92], Swift [24] and Muller [34] present the esti-

TABLE 17.1. EXCESS COMPLIANCE, STIFFNESS AND k/K RATIOS OF DIFFERENT SHEAR JOINTS FOR $P_1/D = 5$, $D/t = 4$, AND $P_2/P_1 = 1$

JOINT / HEAD	σ MPa	P_1/D_s	D_s/t	Method[1]	%Cl	%X	C M/GN	k MN/m	K MN/m	k/K	Reference
Al, butt/std	30	5	2	FE	0	0	16	63	343	0.18	3S-1 (33)
Al, butt/std		5	2.4	M	0	0	13	88	224	0.39	(70)
Al, butt/std	30	5	2	FE	0.5	0	5	200	343	0.58	3S-2 (33)
Steel, butt/std	126	5	2	FE	0.9	0	1.9	526	1013	0.52	3S-3 (1)
Al, lap/std	30	5	4	FE	0	0	35	29	171	0.17	4S-29 (33)
Al, lap/std	30	5	4	FE	0.5	0	25	40	171	0.23	4S-30 (33)
Al, lap/std	126	5	4	FE	0	0	35	29	171	0.17	4S-1 (1)
Steel, butt/std	126	5	4	FE	0	0	12	83	507	0.16	4S-5 (1)
Al, lap/csk	100	5	4	FE	0	0	33	30	172	0.17	4C-1 (1)
Al, lap/csk		5	4 (t=1)	M	0	0	56	18	112	0.16	(43)
Al, lap/csk		5	2.2 (t=2.2)	M	0	0	30	33	246	0.13	(43)
Al, lap/csk		5	2.4	M	0	0	32	31	172	0.18	(70)
Al, lap/csk		5	2	M	0	0	40	25	172	0.15	(70)
Al, lap/csk		5	1.2	M	0	0	21	48	172	0.28	(70)
Al, lap/csk		5	4 (t=1)	M	0	Yes[2]	34	29	112	0.26	(43)
Al, lap/csk		5	2.2 (t=2.2)	M	0	Yes[2]	16	63	246	0.26	(43)

[1] FE-finite element calculation; M-measured

[2] With hole expansion, amount not specified

mates of f_i derived from the 2-spring models for different joints, but they do not document the relations between f_i and the k/K– (or C_P/C–) ratio defined by the models. The expressions for the fastener loads and load fractions for different types of multiple row joints are summarized in Tables 17.2, 17.3 and 17.4. Estimates of the values of f_i for different joints and number of fastener rows are summarized in Table 2.5 in Chapter 2.6.

TABLE 17.2. EXPRESSIONS FOR THE FASTENER LOAD RATIOS, f_i, DERIVED FROM THE 2-SPRING MODEL FOR MULTIPLE ROW LAP AND BUTT JOINTS

NUMBER OF ROWS	PROPERTY	ROW 1	ROW 2	ROW 3	ROW 4
2	FASTENER LOAD, Q	Q_1	$Q_2 = Q_1$		
	FASTENER LOAD FRACTION, f_i	0.5	0.5		
3	FASTENER LOAD, Q	Q_1	$Q_1 [1-k/K]$	$Q_3 = Q_1$	
	FASTENER LOAD FRACTION, f_i	1 / [3-k/K]	[1-k/K] / [3-k/K]	1/[3-k/K]	
4	FASTENER LOAD, Q	Q_1	$Q_1 [1-k/K]$	$Q_1 [1-k/K]$	$Q_4 = Q_1$
	FASTENER LOAD FRACTION, f_i	1 / [4-2k/K]	[1-k/K] / [4-2k/K]	[1-k/K] / [4-2k/K]	1 / [4-2k/K]

* Expressions apply to joints consisting of panels of equal thickness.

TABLE 17.3. EXPRESSIONS FOR THE RATIOS, Q_i / Q_1, AND FASTENER LOAD FRACTIONS, f_i, DERIVED FROM THE 2-SPRING MODEL FOR MULTIPLE ROW ATTACHMENT JOINTS

	Q_i / Q_1			
NUMBER OF ROWS	ROW 1	ROW 2	ROW 3	ROW 4
2	1	$1 / [1+k/K]$		
3	1	A**	$A / [1+k/K]$	
4	1	B***	AB	$AB / [1+k/K]$

*Note: the fastener load fraction, $f = (Q_i /Q_1) / (\Sigma Q_i /Q_1)$

** $A = 1 / \{ 1 + k/K + k / K[1+k/K]\}$

*** $B = 1 / \{1 + k/K+ Ak/K+ Ak/K/ [1-k/K]\}$

TABLE 17.4. EXPRESSIONS FOR THE FASTENER LOAD FRACTIONS, F_i, DERIVED FROM THE 2-SPRING MODEL FOR MULTIPLE ROW DOUBLERS

	FASTENER LOAD FRACTION, fi					
NUMBER OF ROWS	ROW 1	ROW 2	ROW 3	ROW 4	ROW 5	ROW 6
2	$\beta k/K_S$	$\beta k/K_S$				
4	$\alpha[k/K+ f_2]$	$\beta k/K_S[1-f_1]$	$\beta k/K_S[1-f_1]$	$\alpha[k/K+ f_2]$		
6	$\alpha[k/K+ f_2]$	$\alpha[k/K-2kf_1/K+f_3]$	$\beta k/K_S[1-2f_1-2f_2]$	$\beta k/K_S[1-2f_1-2f_2]$	$\alpha[k/K-2kf_1/K+f_3]$	$\alpha[k/K+ f_2]$

$\alpha=[1+2k/K]^{-1}$, $\beta=[1+2k/K_S]^{-1}$

The quantity K is the stiffness of the strip of panel between 2 adjacent fastener rows; K_S is the stiffness of the stri[of panel between the symmetry axis and the adjacent row of fasteners (see Figure 17.1).

17.4 A SUPERPOSITION METHOD FOR ESTIMATING SCFs

With knowledge of the fastener load distribution, the SCFs in a multiple-fastener-row joint can be estimated. A method that superposes the effects of fastener loading and load transmitted to the next fastener row is provided.

Bearing Mode. Panel stresses generated at a particular row of fastener holes by the bearing mode component of the fastener load arise from the superposition of two stress fields. One component is produced by the fastener load, Q_i, the other is produced by the bypass load Q_{BP}, the portion of the load on the panel reaching a particular row that is *not* transferred to the fasteners but remains in the panel and is transmitted across the fastener holes and down the panel to the next row of fasteners:

$$Q_{BP,A} = Q - Q_1, \; Q_{BP,B} = Q - Q_1 - Q_2, \text{ and } Q_{BP,C} = Q - Q_1 - Q_2 - Q_3, \text{etc} \qquad (17.4)$$

where $Q_{BP,A}$, $Q_{BP,B}$ and $Q_{BP,C}$ are the bypass loads for the leading and successive rows of holes, respectively. The bypass load acting on the fastener hole produces a stress field approximated by the field of a filled, open hole panel (see Figure 1.4b) An estimate of the SCF is obtained by summing the SCFs produced by the two components of the field:

1. **Contribution of Fastener Load.** The reductions of the fastener loads acting in multiple row joints are accompanied by reductions in the peak panel stresses. The SCF produced by the fastener load at a particular row i, $SCF_{FL,i}$ is approximated by the SCF of a *comparable* single row joint diminished by the factor, f_i :

$$SCF_{FL,i} = f_i \, (SCF_{SINGLE \, ROW}) \, ... \qquad (17.5)$$

In this context, *comparable* implies that the SCF for the single row joint reflects the same mix of bearing frictional load existing in the multiple row joint.

2. **Contribution of Bypass Load.** An estimate of the $SCF_{BP,i}$, the SCF produced at row i by the bypass load, is obtained by assuming that it is comparable to the SCF of a filled (open) hole panel[1], SCF = 2.5, diminished by $f_{BP,i} = Q_{BP,i}/Q$, the fraction of the total load that bypasses that row.

$$SCF_{BP,i} = f_{BP,i} \, (SCF_{FILLED \, HOLE}) = 2.5 \, f_{BP,i} \, ... \qquad (17.6)$$

where:

$$f_{FB,A} = 1 - f_1, \; f_{FB,B} = 1 - f_1 - f_2, \text{ etc.} \, ... \qquad (17.7)$$

The SCF at a particular row of a multiple row joint, SCF_i, is the sum of these 2 contributions:

$$SCF_i = SCF_{FL,i} + SCF_{BP,i} = f_i \, (SCF_{SINGLE \, ROW}) + 2.5 \, f_{BP,i} \, ... \qquad (17.8)$$

The quantity, $SCF_i = \sigma*_i/\sigma$, relates the peak local stress to the nominal (gross section) stress applied to the panel.[2] Estimates of the SCF of different multiple row joints derived from the 2-spring model and the superposition method are listed in Table 2.5.

Clamped Mode. Clamping shields the edge of the fastener hole from stress. The finite element calculations show that the SCFs for a fully clamped, single fastener row and double-row butt joint, $P_1/D_S = 5$, $D_S/t = 4$, (Table 3.4, Chapter 3.1) are the same. According to the superposition method, this means that relevant by-pass SCF for the clamping mode is

---

[1] Earlier findings for a narrow, single fastener butt joint with free edges showed that the SCF of a filled hole panel is 2.85 at 22° and 158°, 2.1 at 0° and 180°. Since the peak values of single row shear joints generally fall between these 2 locations, the SCF of the filled hole was approximated by the average value, SCF = 2.5. This value was used to estimate the SCFs listed in Tables 2.5 and 17.5. More recent findings for a wide butt joint (see Figure 17.2) show the filled-hole SCF = 2.6 at 0° and 180°. The authors believe the 4% difference is within the uncertainty of the estimation procedure.

[2] It should be noted that the reductions of the SCFs of multiple row joints arise from the reductions of the fastener loads rather than a reduction of the stress concentrating power of the fastener-hole configuration.

$SCF_{BP} \approx 1.8$ compared to $SCF_{BP} \approx 2.5$ for the bearing mode. The implication is that the benefits of multiple fastener rows are meager for clamped butt because their single row SCF, (SCF = 1.8 for $P_1/D_S = 5$, $D_S/t = 4$) is close to the by-pass value. The bearing mode SCF is dominant when load transfer proceeds by a combination of the bearing and clamping mode. In that case, estimates of the bearing mode SCF can be obtained following the methods described above after reducing the fastener load by the amount supported by clamping.

Hole Expansion. Estimates of the fastener and by-pass loads of multi row joints fabricated with hole expansion can be derived from the 2-springs models with the appropriate stiffness ratio. These loads may ultimately be helpful for predicting the strength reduction factor (SRF) of multi row joints (see Chapter 4.3). The analytical approach presents problems because the SCF is not a valid descriptor of the elastic-plastic, stress-strain states of joints with expanded holes. Alternative descriptors, such as the peak plastic strain amplitude, are difficult to evaluate with existing finite element capabilities. The experimental approach, correlation of the SRF with number of fastener rows, is obviated by the absence of data for hole expanded joints.

17.5 VALIDATION

Load distributions derived from the 2-spring model are validated by finite element calculations and estimates from other analyses by Swift [24] and others [92-94] as shown in Tables 17.5 and 17.6. The finite element calculations of the SCFs of multiple row attachment joints in Table 17.5 are also in good agreement with the values obtained with the superposition method.

TABLE 17.5. COMPARISONS OF ESTIMATES OF THE FASTENER LOAD FRACTIONS, f_i, AND SCFS FOR MULTIPLE ROW ATTACHMENT JOINTS DERIVED FROM THE 2-SPRING MODEL, FOR k/K = 0.5, AND THE SUPERPOSITION METHOD WITH FINITE ELEMENT CALCULATIONS.

NUMBER OF ROWS	SOURCE	ROW 1		ROW 2		ROW 3		ROW 4	
		f_i	SCF	f_i	SCF	f_i	SCF	f_i	SCF
1	(2P-1)	1	**6.2**						
2	2 Spring Model + Superposition (2P-2)	0.60	**4.7**	0.40	**2.5**				
	FEA	0.60	**4.7**	0.40	**2.4**				
3	2 Spring Model + Superposition (2P-3)	0.52	**4.4**	0.29	**2.3**	0.19	**1.2**		
	FEA	0.52	**4.5**	0.29	**2.2**	0.19	**1.2**		
4	2 Spring Model + Superposition (2P-4)	0.54	**4.5**	0.24	**2.0**	0.13	**1.0**	0.09	**0.06**
	FEA	0.50	**4.4**	0.26	**2.2**	0.15	**1.1**	0.10	**0.06**

* Aluminum joint with $P_i/D = 4$, $D/t = 5$, and $P_2/P_1 = 1$.

**TABLE 17.6. COMPARISONS OF ESTIMATES OF THE FASTENER
LOAD FRACTIONS, f_i, FOR MULTIPLE ROW JOINTS DERIVED
FROM THE 2-SPRING MODEL WITH PREVIOUSLY
REPORTED VALUES**

JOINT TYPE	FASTENER LOAD FRACTION, f_i	
	2-Springs Model	Reported Value* (Reference)
3 rivet-row lap joint		
leading or lagging row	0.36	0.35 (92), 0.36 (95), 0.375 (24)
intermediate row	0.28	0.28 (95), 0.295 (92)
4 rivet-row lap joint		
leading or lagging row	0.31, 0.28	0.32 (24)
4 rivet-row doubler		
leading or lagging row	0.19	0.14 (92)
intermediate rows	0.06	0.04 (92)

*Note: The reported values are for similar but not identical panel and fastener dimensions.

ANALYSIS OF FASTENER TENSION AND CLAMPING STRAIN

18.1 NOMENCLATURE

E_F = fastener material elastic modulus
E_P = panel material elastic modulus
A_S = fastener shank cross-sectional area. $A_S = \pi D_S^2/4$
A_P = cross-sectional area of panels under the fastener head. $A_P = \pi[D_B^2 - D_S^2]/4$
A_H = fastener head cross-sectional area. $A_H = A_P$
H_H = fastener head height
H_0 = fastener shank height
$2t$ = combined thickness of panels. $2t \approx H_0$
k_1 = Fastener shank stiffness. $k_1 = A_S E_F/H_0$
k_2 = Combined panels stiffness. $k_2 = A_P E_P/2t$

$$k_2^{eff} = \left[\frac{1}{k_2} + \frac{2}{k_3} \right]^{-1}$$

k_3 = Fastener head stiffness. $k_3 = A_H E_F/H_H$

$$k^* = \frac{k_2^{eff}}{k_1 + k_2^{eff}} = \frac{Q_c}{Q_n}$$

x – nominal fastener shank misfit $x = H_0 \varepsilon_{N,cl}$
x' = residual shank extension
x'' = combined panels compression
Q_N = tension applied to shank to produce nominal shank extension. $Q_N = k_1 x$
Q_C = residual tension in fastener shank. $Q_C = k_1 x'$
$\varepsilon_{N,cl}$ = nominal clamping strain. $\varepsilon_{N,cl} = \%CL/100$
ε = residual shank strain

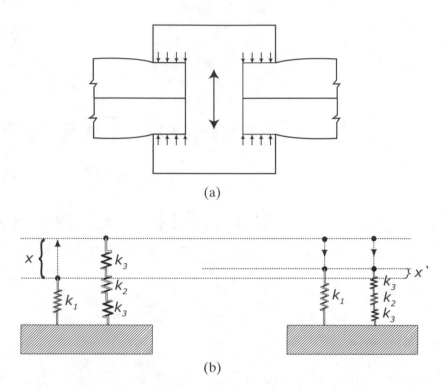

(a)

(b)

Figure 18.1. Elastic-loading-unloading model of the clamping process using linear springs k_1, k_2 and k_3 to represent the fastener shank, panels and fastener head, respectively. The initial shank height misfit, x, is overcome by tensile loading. Unloading the fastener shank generates compression in the panels and fastener heads.

$\sigma_{N,cl}$ = nominal clamping stress. $\sigma_{N,cl} = E_s \varepsilon_{N,cl} = \dfrac{Q_N}{A_s}$

σ_S = residual tensile stress in fastener shank. $\sigma_S = E_S \varepsilon = \dfrac{Q_N}{A_s}$

18.2 CLAMPING MODEL

Closed-form expressions for the average fastener tension and clamping strain are derived. The equations relate average clamping state parameters to fastener and panel material properties, thicknesses and diameters. They account for the compliance of the panels and non-uniform fastener head deformation.

The Model. The generation of the clamped condition in a shear joint can be modeled as a 2-step process using a system of linear springs arranged in series and parallel, as shown in

Figure 18.1. The first step is the tensile loading of the fastener shank (minus fastener heads) until the nominal value of the axial (clamping) strain is reached. The second step is the transfer of load to the panels and the two fastener heads which are compressed allowing the fastener shank to partially relax.

Affects of Panel Compliance and Non-Uniform Fastener Head Deformation. The fastener shank is first loaded in tension to ($Q_N = k_1 x$) to overcome the initial misfit, x. The portions of the panels outside the boundaries of the fastener heads, and frictional slip between the fastener/head and panel/panel interfaces are neglected. The following equilibrium and deformation compatibility conditions must be satisfied after relaxation (see Figure 18.1(b)):

$$\text{Equilibrium:} \quad Q_C = k_1 x' = x'' \tag{18.1}$$
$$\text{Compatibility:} \quad x - x' = x'' \tag{18.2}$$

After relaxation, the resultant load is shared by the panels, the fastener shank and the fastener head. Substituting for x'' from (18.2) in (18.1) and noting $2t \approx H_0$, we get

$$x' = \left(\frac{k_2^{eff}}{k_1 + k_2^{eff}} \right) x = K^* x \tag{18.3}$$

$$\text{where} \quad k_2^{eff} = \left(\frac{1}{k_3} + \frac{1}{k_2} + \frac{1}{k_3} \right)^{-1} = \left(\frac{2}{k_3} + \frac{1}{k_2} \right)^{-1} \tag{18.4}$$

$$\text{and} \quad K^* = \left(\frac{k_2^{eff}}{k_1 + k_2^{eff}} \right) = \frac{x'}{x} = \frac{\varepsilon}{\varepsilon_{N,cl}} = \frac{\sigma_S}{\sigma_{N,cl}} = \frac{Q_C}{Q_N} \tag{18.5}$$

Deformation of the fastener head is non-uniform. Since most of it is distributed across the annular region that surrounds the shank diameter i.e., the ring of fastener head material that directly overlaps the panels, only this portion of the fastener heads is considered, i.e., $k_3 = A_H E_F / H_H = A_P E_F / H_H$.

18.3 FASTENER TENSION AND CLAMPING STRAIN PREDICTIONS

Influence of Panel Compliance. The equations derived above can be applied to a typical aluminum alloy lap or butt joint considered in this book: $E_F = E_P = 70$ GPa, $D_S = 6.12$ mm, $H_0 \approx 2t = 3.06$ mm, $D_B = 9.792$ mm. Setting $k_3 = 0$, This gives,

$$k_1 = 672.93 \text{ kN / mm}, \ k_2 = 1049.8 \text{ kN / mm, and } K^* = 0.61 = \frac{x'}{x} = \frac{\varepsilon}{\varepsilon_{N,cl}} = \frac{\sigma_S}{\sigma_{N,cl}}$$

This means that the clamping strain in this case is only 61% of the nominal value (assuming a rigid fastener head). Alternately, 39% of the nominal clamping strain is lost due to compressive deformation of the panels underneath the fastener head. For the prototypical lap joint considered in this book with a shank height of 3.06 mm and a nominal clamping strain

$\varepsilon_{N,cl} = 0.005$ (0.5%), the corresponding value of shank extension is $x = 0.0153$ mm, and Equation (18.9) can be used to define the effective clamping state parameters as follows:

ε:	0.0031 (0.31% clamping)
x':	0.0093 mm
σ_S:	213.3 MPa
panels deformation:	0.0060 mm

The average fastener shank stress due to clamping is 61% of the nominal value, $\sigma_N = 350$ MPa. Using Equations (18.4) and (18.9), it is possible to estimate the influence of different material combinations on the effective clamping. If the fastener material is steel ($E_F = 206$ GPa) instead of aluminum, $K^* = 0.35$ and the effective clamping strain would be only 0.18% for a nominal clamping strain of 0.5%. Despite the reduction in K^*, the steel fastener shank stress, $\sigma_S = 356.8$ MPa, is still higher than with an aluminum fastener. This distinction between K^* and the actual clamping stress, σ_S, is an important one and must be kept in mind when interpreting K^*.

Influence of Non-Uniform Fastener Head Deformation. For the prototypical lap joint considered in this book, $H_H = 3.83$ mm. This gives $k_1 = 838.72$ kN/mm which

K^*:	0.31
ε:	0.00155 (0.155% clamping)
$x^{`}$:	0.004743 mm
σ_S:	107.8 MPa
$Q_C = \sigma_S A_S$	3189.9 N
panels deformation:	($Q_C/k_2 =$) 0.003038 mm
fastener heads deformation:	($2 Q_C/k_3$) = 0.007519 mm

means that the shank strain is only about 0.16% for a nominal clamping of 0.5%; compressive deformation of the panels and fastener head cause a loss of about 20% and 49%, respectively, of the nominal clamping strain. With a steel fastener, $K^* = 0.22$, the shank strain is 0.11%. The steel fastener shank stress, $\sigma_S = 229.3$ MPa, is higher than with an aluminum fastener. Other examples are presented in Table 3.1.

In Table 18.1, the values of σ_S and Q_S obtained with this analysis for several joints (including the one described in the above table) are compared with the average tension at the shank midsection and corresponding Q_S derived from the finite element analyses. The analytical values for the relatively short rivets (first three entries) understate the more reliable finite element values by from 10% to 20%. The analytical values overstate the experimentally measured σ_S for the long bolt (4th entry) by 70%. This discrepancy is attributed to significant departures from elastic behavior. The %Cl = 1.3 clamping caused the bolt to undergo plastic deformation. The authors believe this both increases the accommodation at the bolt head and reduces the elastic stretch of the shank, features not accounted for by the analysis.

TABLE 18.1. DISTRIBUTION OF ACCOMMODATIONS TO THE INITIAL CLAMPING MISFIT

MATERIAL	JOINT			CLAMPING		MODEL	k_1	k_2	k_3	K^*	σ_S	Q_C
	H_0	D_S	D_B	%CL	Misfit						FEA[1]/ANAL.[2]/EXP.[3]	FEA[1]/ANAL.[2]/EXP.[3]
	(mm)	(mm)	(mm)		(mm)		(kN/mm)	(kN/mm)	(kN/mm)	(kN/mm)	(MPa)	(kN)
Aluminum	3.06	6.12	9.79	0.5	0.0153	4S-30	672.9	1049.8	838.7	0.31	120/108/--	3.5/3.2/--
Aluminum	6.12	6.12	9.79	0.5	0.0306	3S-3	336.5	524.9	838.7	0.41	170/143/--	5.0/4.2/--
Steel	6.12	6.12	9,79	0,85	0.052	3S-4	990.2	1544.7	2468.2	0.41	854/716/--	25/21/--
Steel [1]	108	22	33	1.3	1.4		725.1	906.3	2966.2	0.44	--/1170/688	--/445/270

[1] ANAL.- Analyses
[2] FEA - Finite element analyses
[3] EXP - Experimental determination for a A354BD bolt (43)

18.4 SUMMARY OF PROCEDURE

The methodology for estimating the average fastener tension, clamping strain and distribution of deformations (compression of the panels, shank and fastener head) can be summarized as follows:

1. Using the fastener elastic modulus (E_F), fastener shank cross-sectional area (A_S), fastener shank nominal height (H_0), panels' elastic modulus (E_P), panel cross-sectional area underneath the fastener head ($A_H = \pi[D_B^2 - D_S^2]/4$) and fastener head height (H_H), obtain stiffness values for the fastener shank (k_1), the panels (k_2) and fastener head (k_3) using

$$k_1 = \text{Fastener shank stiffness} = A_S E_F / H_0$$
$$k_2 = \text{Combined panels stiffness} = A_H E_P / H_0$$
$$k_3 = \text{Fastener head stiffness} = A_H E_F / H_H$$

Note that the initial combined panel thickness ($2t$ for a lap joint) has been approximated with H_0 in the expression for k_2. This approximation simplifies the mathematical procedure without significant loss in accuracy and is reasonable for typical clamping misfit values.

2. Using the equations below, obtain K^*

$$k_2^{eff} = \left(\frac{1}{k_3} + \frac{1}{k_2} + \frac{1}{k_3} \right)^{-1} = \left(\frac{?}{k_3} + \frac{1}{k_2} \right)^{-1}$$

$$K^* = \left(\frac{k_2^{eff}}{k_1 + k_2^{eff}} \right) = \frac{x'}{x} = \frac{\varepsilon}{\varepsilon_{N,cl}} = \frac{\sigma_S}{\sigma_{N,cl}} = \frac{Q_C}{Q_N}$$

Obtain clamping strain, ε, and fastener tension, Q_C.

3. Obtain the deformation of the panels and fastener head as

$$Q_C/k_2 = \text{combined deformation of panels} = x^p$$
$$2Q_C/k_3 = \text{combined deformation of 2 fastener heads} = x^h$$

4. The total deformation is $(x' + x^p + x^h)$ and should equal the nominal clamping misfit (CL). The proportions of deformations are given by

$$x'/CL \text{ (fastener shank)}$$
$$x^p/CL \text{ (panels)}$$
$$x^h/CL \text{ (fastener heads)}$$

SUMMARY OF CALCULATIONAL MODELS

This book presents the results of over 150, 2- and 3-dimensional, finite element analyses of butt, attachment and lap joints. The computations examined six different joint configurations, eight ways of fastening them, and fifteen other geometric, material and loading variables:

JOINT CONFIGURATION

1. Panel with an open hole
2. Attachment joint
3. Butt joint
4. Single fastener row lap joint
5. Double fastener row lap joint
6. Adhesive lap joint
7. Double fastener row doubler

FASTENER TYPE

H. No fastener (open hole)
P. Pin
S. Standard fastener head
C. Countersunk fastener head
A. Adhesive or sealant
SA. Standard fastener combined with adhesive
CA. Countersunk fastener combined with adhesive
SP. Self-Piercing rivet

JOINT VARIABLES

Panel material	Fastener material	Interface friction
Elastic or elastic plastic behavior	Fastener head shape	Uniaxial and biaxial loading
	Interference	
Panel width	Fastener clamping	Adhesive properties
Panel thickness	Number of fastener rows	Adhesive thickness

An alpha-numeric code is used to identify the specific calculational models employed. It consists of the joint configuration number, the fastener-type abbreviation and consecutive numbers for the particular combination of other variables. For example: 4CA-5 is the 5th calculation for a single rivet-row lap joint fastened with both countersunk rivets and adhesive.

The following tables describe all of the finite element calculations referred to in this book. The finite element models used to obtain these data are described in detail in Chapter 13; the material properties for the various models are described in Chapter 15.

TABLE A.1. SUMMARY OF 2-D CALCULATIONAL MODELS FOR OPEN HOLE PANELS: INFINITE PLATES WITH A SINGLE, CENTRALLY LOCATED EMPTY HOLE, UNLESS OTHERWISE NOTED. NOMINAL STRESS 125 MPa, WITH NO FRICTION.

Model No	Description of Holes	Reference
1H-1	Straight sided hole	Iyer [1]
1H-2	61.6 degree countersunk hole to full panel thickness	Iyer [1]
1H-3	100 degree countersunk hole to half panel thickness	Iyer [1]
1H-4	100 degree countersunk hole to full panel thickness	Iyer [1]
1H-5	Straight sided filled hole	Iyer [35]

TABLE A.2. SUMMARY OF 2-D CALCULATIONAL MODELS FOR INFINITELY WIDE, MULTIPLE-ROW, ALUMINUM ATTACHMENT JOINTS WITH HEADLESS FASTENERS, $P_1/D = 5$, $P_2/D = 5$, LINEAR ELASTIC BEHAVIOR AND UNIAXIAL LOADING; NOMINAL STRESS 125 MPa, AND COEFFICIENT OF FRICTION OF 0.2.

Model No.	Number of Rows	Refererence
2P-1	1	Iyer et al. [35]
2P-2	2	Iyer et al. [35]
2P-3	3	Iyer et al. [35]
2P-4	4	Iyer et al. [35]

TABLE A.3. SUMMARY OF CALCULATIONAL MODELS FOR BUTT JOINTS; UNLESS OTHERWISE NOTED, THESE ARE FOR 2-D INFINITE PANELS WITH A SINGLE HOLE, HEADLESS FASTENER, LINEAR ELASTIC BEHAVIOR AND UNIAXIAL TENSION LOADING (σ_{NOM} = 60 MPa).

Model No.	Pin-Panel Materials	Coefficient Friction	Special Features	Reference
3P-1	Al-Al	0.2	Row of pinned holes, $P_1/D=5$, $\sigma_{nom}=125$ MPa	Iyer [1]
3P-2	Al-Al	0		Iyer [18]
3P-3	Steel-Al	0		Iyer [18]
3P-4	Ti-Al	0		Iyer [18]
3P-5	Al-Epoxy	0		Iyer [18]
3P-6	Al-Al	0.2		Iyer [18]
3P-7	Steel-Al	0.2		Iyer [18]
3P-8	Ti-Al	0.2		Iyer [18]
3P-9	Al-Epoxy	0.2		Iyer [18]
3P-10	Al-Al	1		Iyer [18]
3P-11	Steel-Al	1		Iyer [18]
3P-12	Ti-Al	1		Iyer [18]
3P-13	Al-Al	0	Finite width, $L_1=30.6$ mm, $L_1/D=5$	Iyer [18]
3P-14	Steel-Al	0	Finite width, $L_1=30.6$ mm, $L_1/D=5$	Iyer [18]
3P-15	Ti-Al	0	Finite width, $L_1=30.6$ mm, $L_1/D=5$	Iyer [18]
3P-16	Al-Al	0.2	Finite width, $L_1=30.6$ mm, $L_1/D=5$	Iyer [18]
3P-17	Steel-Al	0.2	Finite width, $L_1=30.6$ mm, $L_1/D=5$	Iyer [18]
3P-18	Al-Al	1	Finite width, $L_1=30.6$ mm, $L_1/D=5$	Iyer [18]
3P-19	Al-Al	0	Row of pinned holes, $P_1/D=5$	Iyer [18]
3P-20	Steel-Al	0	Row of pinned holes, $P_1/D=5$	Iyer [18]
3P-21	Al-Al	0.2	Row of pinned holes, $P_1/D=5$	Iyer [18]
3P-22	Steel-Al	0.2	Row of pinned holes, $P_1/D=5$	Iyer [18]
3S-1	Al-Al	0.0, 0.2	3-D analysis, single row of standard head fasteners, $P_1/D=5$, $0 \leq \sigma \leq 150$	Iyer [35]
3S-2	Steel-Steel	0.2	3-D analysis, single row of standard head fasteners, %Cl=0.85, $P_1/D=5$, $0 \leq \sigma \leq 375$	Iyer [35]
3S-3	Al-Al	0.0, 0.2	3-D analysis, single row of standard head fasteners, %Cl=0.5, $P_1/D=5$, $0 \leq \sigma \leq 150$	Iyer [35]
3S-4	Steel-Steel	0.2	3-D analysis, double row of standard head fasteners, %Cl=0.85, $P_1/D=5$, $0 \leq \sigma \leq 375$	Iyer [35]

TABLE A.4. SUMMARY OF 3-D CALCULATIONAL MODELS FOR SINGLE RIVET-ROW LAP JOINTS; UNLESS OTHERWISE NOTED, THESE ARE FOR STANDARD RIVET HEADS, INFINITELY WIDE LAP JOINTS, $P_1/D = 5$, $D/t = 4$, ALUMINUM RIVETS AND PANELS, UNIAXIAL LOADING, LINEAR ELASTIC BEHAVIOR AND COEFFICIENT OF FRICTION OF 0.2.

Model No.	Nominal Stress MPa	Other Special Features	Reference
4S-1	125		Iyer et al. [1,13,15]
4S-2	125	Coef. friction = 0.5	Iyer et al. [1,13,15]
4S-3	125	Elastic-isotropic-plastic	Iyer et al. [1,13,15]
4S-4	125	Updated ABAQUS	Loha [3]
4S-5	125	Steel rivets and plates, updated ABAQUS	Loha [3]
4S-6	62.6		Dechwayukul et al. [6,21]
4S-7	62.6	Finite width, L = 30.6 mm	Dechwayukul et al. [6,21]
4S-8	65		Fongsamootr [7]
4S-9	65	Finite width, L = 30.6 mm	Fongsamootr [7]
4S-10	65	D/t = 6.12	Fongsamootr [7]
4S-11	65	Finite width, L = 30.6 mm, D/t = 6.12	Fongsamootr [7]
4S-12	100	Elastic-isotropic-plastic	Al Dakkan [4]
4S-13	100	Elastic-isotropic-plastic, no rivet head	Al Dakkan [4]
4S-14	100	Elastic-isotropic-plastic, finite width, L = 30.6mm	Al Dakkan [4]
4S-15	100	Elastic-isotropic-plastic, finite width, L = 30.6mm and no rivet head	Al Dakkan [4]
4SP-1	55.6	Self-piercing rivets	Iyer et al. [44]
4SP-2	55.6	Self-piercing rivets	Iyer et al. [44]
4SP-3	55.6	Self-piercing rivets	Iyer et al. [44]
4S-29	0-150	μ = 0, 0.2, 0.4	Iyer [35]
4S-30	0-150	μ = 0, 0.2, 0.4, %Cl = 0.5	Iyer [35]

TABLE A.5. SUMMARY OF 3-D CALCULATIONAL MODELS FOR SINGLE RIVET- ROW LAP JOINTS WITH COUNTERSUNK RIVETS; UNLESS OTHERWISE NOTED, THESE ARE INFINITELY WIDE JOINTS, $P_1/D = 5$. $D/t = 4$, ALUMINUM RIVETS AND PANELS, LINEAR ELASTIC BEHAVIOR, 100° COUNTERSUNK TO HALF PANEL THICKNESS, UNIAXIAL LOADING, WITH A COEFFICIENT OF FRICTION OF 0.2.

Model No.	Nominal Stress	Other Special Features	Reference
4C-1	125		Iyer [1,13,15]
4C-2	125	Steel rivets	Iyer [1,13,15]
4C-3	125	Elastic-isotropic-plastic	Iyer [1,13,16]
4C-4	125	Coef. Friction = 0.4	Iyer [1,13,15]
4C-5	125	61.6° head c.s. to full thickness	Iyer [1,13,15]
4C-6	125	100° head c.s. to full thickenss	Iyer [1,13,15]
4C-7	62.6		Dechwayukul [6]
4C-8	62.6	Finite width, L = 30.6 mm	Dechwayukul [6]
4C-9	60		Kamnerdtong [8]
4C-10	60	Finite width, L = 30.6 mm	Kamnerdtong [8]
4C-11	60	D/t = 6.12	Kamnerdtong [8]
4C-12	60	Finite width, L = 30.6 mm, D/t = 6.12	Kamnerdtong [8]
4C-13	100	Elastic-isotropic-plastic	Huang [5]
4C-14	100	Elastic-isotropic-plastic, no rivet head	Huang [5]
4C-15	100	Elastic-isotropic-plastic, finite width, L = 30.6mm	Huang [5]
4C-16	100	Elastic-isotropic-plastic, finite width, no head	Huang [5]

**TABLE A.6. 3-D COMPUTATIONAL MODELS FOR
DOUBLE RIVET-ROW LAP JOINTS; UNLESS OTHERWISE
NOTED, THESE ARE FOR INFINITELY WIDE JOINTS,
$P_1/D = 5$, $D/t = 4$, ALUMINUM RIVETS AND PANELS,
LINEAR ELASTIC BEHAVIOR, UNIAXIAL LOADING,
A NOMINAL STRESS OF 125 MPa, AND A COEFFICIENT
OF FRICTION OF 0.2.**

Model No.	Other Special Features	Reference
5S-1	Standard rivet heads	Iyer [1,13,19]
5C-1	100 degree head countersunk to 1/2 panel thickness	Iyer [1,13,20]
5C-2	100 degree head countersunk to 1/2 panel thickness, steel rivet	Iyer [1,13,20]
5C-3	100 degree head countersunk to 1/2 panel thickness, ELKP	Iyer [35]
3S-4	Double row butt joint (see Tables A.3 and A.7)	

TABLE A.7. COMPUTATIONAL MODELS THAT EVALUATE THE EFFECTS OF INTERFERENCE AND CLAMPING; UNLESS OTHERWISE NOTED, THESE ARE INFINITELY WIDE ALUMINUM PANELS WITH ROWS OF ALUMINUM FASTENERS, COEFFICIENT OF FRICTION OF 0.2 AND NOMINAL STRESS OF 125 MPa. MODELS 25-39 ARE 2-D; ALL OTHERS ARE 3-D.

Model No.	Joint type	Matl. Behav.	Coef. Frict.	% Interfer.	%Clamp.	Ref.
3P-25	Butt joint fastened with pins	EIP	0.2	0	0	Iyer [1]
3P-26	Butt joint fastened with pins	EIP	0.2	1	0	Iyer [1]
3P-27	Butt joint fastened with pins	EIP	0.2	2	0	Iyer [1]
3P-28	Butt joint fastened with pins	EIP	0.5	0	0	Iyer [1]
3P-29	Butt joint fastened with pins	EIP	0.5	0.5	0	Iyer [1]
3P-30	Butt joint fastened with pins	EIP	0.5	0.75	0	Iyer [1]
3P-31	Butt joint fastened with pins	EIP	0.5	1	0	Iyer [1]
3P-32	Butt joint fastened with pins	EIP	0.5	2	0	Iyer [1]
3P-33	Butt joint fastened with pins	ELKP	0	0	0	Iyer [1]
3P-34	Butt joint fastened with pins	ELKP	0.2	0	0	Iyer [1]
3P-35	Butt joint fastened with pins	ELKP	0.3	0	0	Iyer [1]
3P-36	Butt joint fastened with pins	ELKP	0.5	0	0	Iyer [1]
3P-37	Butt joint fastened with pins	ELKP	0.2	0.5	0	Iyer [1]
3P-38	Butt joint fastened with pins	ELKP	0.2	0	0	Iyer [1]
3P-39	Butt joint fastened with pins	ELKP	0.2	0.5	0	Iyer [1]
3P-40	Butt joint fastened with pins	Elastic	0.2	0, 0.3, 0.6	0	Iyer [35]
3P-41	2024-T3 Al (σ_Y = 324 MPa) butt joint fastened with pins	EIP	0.2	0.6	0	Iyer [35]
4S-19	1 rivet-row lap joint, std head	ELKP	0.4	0	0	Iyer [1]
4S-20	1 rivet-row lap joint, std head	ELKP	0.4	1	0	Iyer [1]
4S-22	1 rivet-row lap joint, std head	ELKP	0.4	0	0.5	Iyer [1]
4S-23	1 rivet-row lap joint, std head	ELKP	0.4	1	0.5	Iyer [1]
4C-24	1 rivet-row lap joint, CS head	ELKP	0.4	0	0	Iyer [1]
4C-25	1 rivet-row lap joint, CS head	ELKP	0.4	1	0	Iyer [1]
4C-27	1 rivet-row lap joint, CS head	ELKP	0.4	0	0.5	Iyer [1]
5C-3	2 rivet-row lap joint, CS head	ELKP	0.4	0	0	Iyer [1]
5C-4	2 rivet-row lap joint, CS head	ELKP	0.4	1	0	Iyer [1]
3S-1	1 rivet-row butt joint, standard head, $0 \leq \sigma \leq 150$ MPa	Elastic	0.0, 0.2	0	0	Iyer [35]
3S-2	1 rivet-row butt joint, standard head, $0 \leq \sigma \leq 375$ MPa	Elastic, Steel	0.2	0	0.85	Iyer [35]
3S-3	1 rivet-row butt joint, standard head, $0 \leq \sigma \leq 150$ MPa	Elastic	0.0, 0.2	0	0.5	Iyer [35]
3S-4	2 rivet-row butt joint, standard head, $0 \leq \sigma \leq 375$ MPa	Elastic, Steel	0.2	0	0.85	Iyer [35]
4S-29	1 rivet-row lap joint, standard head	Elastic	0.0, 0.2, 0.4	0	0	Iyer [35]
4S-30	1 rivet-row lap joint, standard head	Elastic	0.0, 0.2, 0.4	0	0.5	Iyer [35]
4S-31	1 rivet-row lap joint, standard head	Elastic-Plastic	0.2	1	0	Iyer [35]
7S-1	2 rivet-row doubler, standard head	Elastic	0.2	0	0	Iyer [35]

**TABLE A.8. COMPUTATIONAL MODELS THAT EVALUATE THE EFFECTS
OF BIAXIAL LOADING; UNLESS OTHERWISE NOTED THESE ARE
FOR INFINITELY WIDE, 2-D ALUMINUM PANELS OF A BUTT JOINT
WITH ROWS OF FASTENERS, LINEAR ELASTIC BEHAVIOR,
AND A COEFFICIENT OF FRICTION OF 0.2, $\mu = 0.31$.**

Model No.	Joint Type	Trans.-to-Long. Stress Ratio	Special Features
3P-2	Single pin	0	Coef. friction = 0
3P-3	Single pin	0	Coef. friction = 0, steel pin ($\mu = 0.33$)
3P-6	Single pin	0	Coef. friction = 0, titanium pin
3P-7	Single pin	0	Coef. friction = 0.2
3P-18	Row of pins	0.33	Coef. friction = 0
3P-19	Row of pins	0.33	Coef. friction = 0, steel pin ($\mu = 0.33$)
3P-20	Row of pins	0.33	Coef. friction = 0, titanium pin
3P-21	Row of pins	0.33	Steel pin
3P-22	Row of pins	0.25	
3P-23	Row of pins	0.5	
3P-24	Row of pins	0.75	
4S-16	Lap joint	0.25	Wide joint, std. head rivets
4S-17	Lap joint	0.5	Wide joint, std. head rivets
4S-18	Lap joint	0.75	Wide joint, std. head rivets
4C-17	Lap joint	0.25	Wide joint, CS rivets
4C-18	Lap joint	0.5	Wide joint, CS rivets
4C-19	Lap joint	0.75	Wide joint, CS rivets

TABLE A.9. 3-D COMPUTATIONAL MODELS FOR EVALUATING THE EFFECTS OF SEALANTS AND ADHESIVES; UNLESS OTHERWISE NOTED THESE ARE FOR INFINITELY WIDE, 1.53 mm-THICK, ALUMINUM, SINGLE RIVET-ROW, LAP JOINTS FASTENED WITH ALUMINUM RIVETS, $P_1/D = 5$, $D/t = 4$, AND A 180 MICROMETER-THICK LAYER OF PRC-DESOTO PR-1776 B-2 (654) SEALANT. SOME MODEL 1 mm-THICK PANELS, $P_1/D = 5$, $D/t = 6.12$. THE MODELS TREAT LINEAR ELASTIC BEHAVIOR.

Model No.	Nom. Stress Mpa	Special Features
4S-6	62.7	Std. rivet head, no sealant
4SA-1	62.7	Std. rivet head
4S-7	62.7	Finite width, std. head, no sealant
4SA-2	62.7	Finite width, std. head
4C-7	62.7	Countersunk head, no sealant
4CA-3	62.7	Countersunk head
4C-8	62.7	Finite width, countersunk, no sealant
4CA-4	62.7	Finite width, countersunk
4S-8	65	Std. rivet head, no sealant
4SA-5	65	Std. rivet head
4S-9	65	Finite width, std. head, no sealant
4SA-6	65	Finite width, std. head
4S-10	65	1 mm-thick panel, std. head, no sealant
4SA-7	65	1 mm-thick panel, std. head
4S-11	65	1 mm-thick, finite width, std. head, no sealant
4SA-8	65	1 mm-thick, finite width, std. head,
4SA-9	65	t (seal) = 150 um, finite, std. head
4SA-10	65	t (seal) = 191 um, finite, std. head
4SA-11	65	t (seal) = 256 um, finite, std. head
4SA-12	65	t (seal) = 382 um, finite, std. head
4SA-13	65	t (seal) = 765 um, finite, std. head
4SA-14	65	t (seal) = 150 um,1 mm-panel finite, std. head
4SA-15	65	t (seal) = 191 um,1 mm-panel, finite, std. head
4SA-16	65	t (seal) = 256 um,1 mm-panel, finite, std. head
4SA-17	65	t (seal) = 382 um,1 mm-panel, finite, std. head
4SA-18	65	t (seal) = 765 um,1 mm-panel, finite, std. head
4SA-6	65	E (seal) = 1.1 Mpa, finite, std.head
4SA-19	65	E (seal) = 4.4 Mpa, finite, std.head
4SA-20	65	E (seal) = 17.6 Mpa, finite, std.head

(table continues)

TABLE A.9. *(CONTINUED)*

Model No.	Nom. Stress MPa	Special Features
4SA-21	65	E (seal) = 70.4 Mpa, finite, std.head
4SA-8	65	E (seal) = 1.1 Mpa, 1 mm-panel, finite, std.head
4SA-22	65	E (seal) = 4.4 Mpa, 1 mm-panel, finite, std.head
4SA-23	65	E (seal) = 17.6 Mpa, 1 mm-panel, finite, std.head
4SA-24	65	E (seal) = 70.4 Mpa, 1 mm-panel, finite, std.head
4C-9	60	CS rivet head, no sealant
4CA-25	60	CS rivet head
4C-10	60	Finite width, CS rivet head, no sealant
4CA-26	60	Finite width, CS rivet head
4C-11	60	1 mm-thick panel, CS head, no sealant
4CA-27	60	1 mm-thick panel, CS head
4C-12	60	1 mm-thick, finite width, CS head, no sealant
4CA-28	60	1 mm-thick, finite width, CS head
4CA-29	60	t (seal) = 150 um, finite width, CS head
4CA-30	60	t (seal) = 191 um, finite width, CS head
4CA-31	60	t (seal) = 256 um, finite width, CS head
4CA-32	60	t (seal) = 382 um, finite width, CS head
4CA-33	60	t (seal) = 765 um, finite width, CS head
4CA-34	60	t (seal) = 150 um,1 mm-panel, finite width, CS head
4CA-35	60	t (seal) = 191 um,1 mm-panel, finite width, CS head
4CA-36	60	t (seal) = 256 um,1 mm-panel, finite width, CS head
4CA-37	60	t (seal) = 382 um,1 mm-panel, finite width, CS head
4CA-38	60	t (seal) = 765 um, 1 mm-panel, finite width, CS head
4CA-6	60	E (seal) = 1.1 MPa, finite width, CS head
4CA-39	60	E (seal) = 4.4 MPa, finite width, CS head
4CA-40	60	E (seal) = 17.6 MPa, finite width, CS head
4CA-41	60	E (seal) = 70.4 MPa, finite width, CS head
4CA--8	60	E (seal) = 1.1 MPa,1 mm-panel, finite, CS head
4CA-42	60	E (seal) = 4.4 MPa, finite width, CS head
4CA-43	60	E (seal) = 17.6 MPa, finite width, CS head
4CA-44	60	E (seal) = 70.4 MPa, finite width, CS head
6A-1	3.8, 12.3, 17.6	Finite width,t(panel) = 1.6mm, t (seal) = 125um,
6A-2	14.4, 52.7, 78.4	Stepped lap joint, finite width, t (panel) = 2mm, t (seal) = 125 um

STRESS CONCENTRATION FACTOR-FATIGUE STRENGTH (SCF-FS) ANALYSIS

A number of workers have demonstrated that the S-N curve (stress amplitude-cycles to failure relation) of a notched component is approximated by the S-N curve of the material divided by the SCF of the notch [65,66]. This is the basis for an estimation procedure, which the authors refer to here as the *Stress Concentration Factor-Fatigue Strength (SCF-FS) Analysis*. The analysis makes it possible to estimate the fatigue strength of components with stress concentrations when: (a) the S-N curve for the material (unnotched samples) and (b) the SCF of the component are known. The following paragraphs identify the limitations of the SCF-FS analysis, provide validation, and demonstrate the analysis can be applied to shear joints.

For convenience, the crack initiation life is arbitrarily defined here as the number of cycles consumed by the nucleation of the crack and its growth to 1 mm-length.[1] Measurements of the initiation and growth lives of lap joints in Table B.1 demonstrate that the initiation life usually dominates [34,55]. The dominance of the initiation life and the fact that the S-N curves of shear joints display very small negative slopes for large values of the cyclic life (see Figures 4.3 through 4.6, 4.9 and 4.10), have 2 important implications. First, the initiation life approximates the total life. Secondly, for long lives, i.e. $N > 10^6$ cycles, a useful and conservative estimate of the stress amplitude for a given *total* fatigue life is obtained from the stress amplitude for the same initiation life. The initiation life of a fatigue crack in a notched component is determined by the locally elevated stresses. The SCF may not alter the fatigue process, but merely allows it to proceed more rapidly consistent with the higher stress levels defined by the SCF. These generalizations are supported by the fatigue test data for open hole panels in Figure B.1 [65]. The data illustrate that useful estimates of the total fatigue life of an open hole panel can be derived from the conventional S-N curve of unnotched samples of the material and the SCF of the hole

---

[1] Defined in this way, the initiation life includes the nucleation of the crack and the increment of growth affected by the SCF. It also represents, approximately, the life consumed while the crack is small enough to be hidden by the rivet head.

TABLE B.1. MEASUREMENTS OF THE CRACK INITIATION, N_i, CRACK GROWTH, N_G, AND TOTAL FATIGUE LIVES, N, OF OPEN HOLE PANELS AND RIVETED LAP JOINTS AFTER FAWAZ [55] AND MULLER [34].

TYPE OF LAP JOINT	σ_a MPa	INITIATION LIFE, Ni 10^3 cycles	GROWTH LIFE, NG 10^3 cycles	N_i/N_G	N_i/N
Open hole Alclad 2024-T3 panel	45	103	65	1.6	0.61
100-mm wide, 1.6 mm-thick	45	118	50	3.36	0.70
Open hole Alclad 2024-T3 panel 100-mm wide, 2.0 mm-thick	42	185	54	3.4	0.77
NAS 1097AD6-6 rivets	37.5	129	27	4.8	0.83
96 mm-wide, 2 mm-thick	25	622	239	2.6	0.72
NAS 1097AD5-5 rivets 140 mm-wide, 1.6 mm-thick 3 rivet rows, $P_1=P_2=20$ mm $P_1/D_s=5$	30-70	50-1300	--	5.7-18	0.85-0.95

The identity between Data Set 2A and Curve 1 implies that an open hole does not alter the fatigue process; it merely allows it to proceed more rapidly consistent with the higher stress levels defined by the SCF. The progressive divergence of the SCF-adjusted S-N curve at the higher stress amplitudes begins as the peak stress approaches and exceeds the yield stress and the SCF, an elastic concept, looses its meaning.

The SCF-FS analysis is expected to be reliable when five requirements are satisfied:

1. **Crack Initiation Dominates.** The bulk of the cyclic life must be expended by the early stages of the fatigue process, crack initiation and small amounts of growth before the presence of the crack alters the stress concentration. Contributions of fretting, corrosion, etc., that are not reflected by the S-N curve of the material must be negligible.

2. **Comparable Stressed Volume and Stress State.** The highly stress volume produced by the SCF must be large enough to accommodate the same fatigue process that proceeds in an unnotched coupon without modifying it. This means that the above findings do not apply to sharp notches with high stress concentrations, steep stress gradients or radically altered states of stress.

3. **Restricted Yielding.** The SCF is an elastic concept, and to be meaningful, yielding within the high stress region must be very limited: $\sigma_{max} < 0.9\sigma_Y$ where σ_{max} is the cycle peak stress: $\sigma_{max} = SCF(\sigma_m + \sigma_a)$, and σ_Y, σ_m and σ_a are the yield stress, the mean stress and the stress amplitude, respectively. When $\sigma_{max} > \sigma_Y$, the SCF-FS analysis overstates the stress amplitude and is conservative.

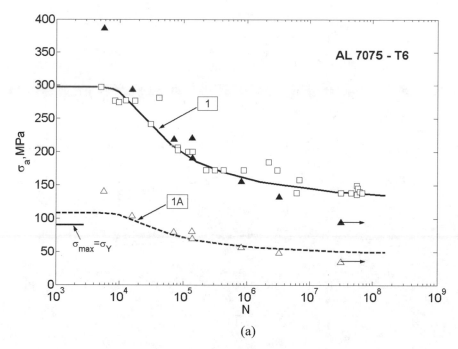

(a)

Figure B.1. Influence of the stress concentration, SCF = 2.76, on the S-N curves of 3 sets of empty hole, aluminum alloy panels [65]: (a) Al 7075-T6, (b) Al 214-T6, and (c) Al 2024-T4. In each case, the stress amplitude versus cyclic life measurements define 2 S-N curves and 3 data sets:

Data Set 1 and Curve 1. Data Set 1 consists of S-N measurements for the aluminum alloy performed on smooth, unnotched test bars, SCF = 1 (open square data points). These measurements define Curve 1.

Data Set 2. Data Set 2 are S-N measurements performed on panels with an empty, open hole (open triangle data points). Note: no curve is drawn for these data.

Curve 1A. The dashed S-N curve, Curve 1A is obtained by dividing the stress amplitudes of Curve 1 (the material S-N curve) by the stress concentration factor of the open hole panel (SCF = 2.76). Curve 1A closely approximates the S-N measurements for the open hole panels (Data Set 2) at all but the highest stress amplitudes.

Data Set 2A. Data Set 2A is obtained by multiplying the nominal stress amplitudes for the open hole panels (Data Set 2) by the SCF (2.76) (solid triangle data points). Data Set 2A is nearly identical to Data Set 1 and is well represented by Curve 1 at all but the highest stress amplitudes. *(figure continues)*

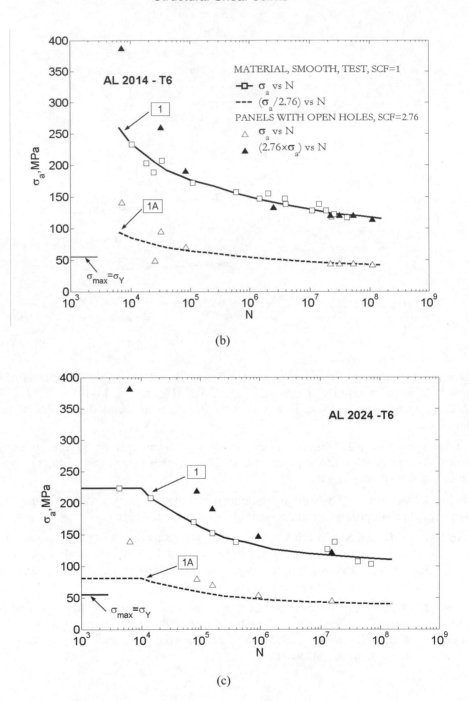

(b)

(c)

Figure B.1. *(continued)*

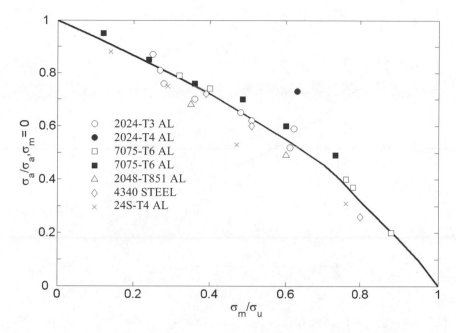

Figure B.2. Gerber diagram constructed with results from a number of sources [92,96-98].

4. **Stress Ratio, R.** The unnotched sample S-N curve of a material and the S-N curve of a notched component are functions of the stress ratio, $R = \sigma_{MIN} / \sigma_{MAX}$. The SCF-FS analysis requires that the unnotched S-N curve corresponds with the stress ratio, R, applied to the notched component. When the material S-N curve for the R-value of interest is not available, an estimate can be derived from the Gerber diagram, shown in Figure B.2, and the relation between R and the mean stress, $\sigma_m = \sigma_a (1 + R)/(1 - R)$. It should be noted that both the stress amplitude and the mean stress appropriate for a notched component are elevated by the stress concentration

5. **S-N Curve.** The S-N curves of structural materials display significant lot- to-lot variation. This is illustrated in Figure B.3. Ideally, the analysis should employ the S-N curve of the same lot of material as the notched component. Since this is usually not possible, uncertainties arising from this must be taken into account.

SCF-FS ANALYSIS FOR SHEAR JOINTS. The SCF-FS analysis applies to shear joints with restrictions. Shear joint panels can generate higher SCF-values than the open hole panel and may also be subject to high mean stresses. For these reasons, Requirement three is only satisfied by relatively low values of nominal stress amplitude, depending on the mean stress, joint configuration and the material's yield stress. Requirement three will also be satisfied at low stress amplitudes where the risk of fretting fatigue is minimal. Requirement two may not be an impediment since, as shown in Figure 2.2, the stress concentration in shear connections are not radically different from that of the open hole panel. Accordingly, the

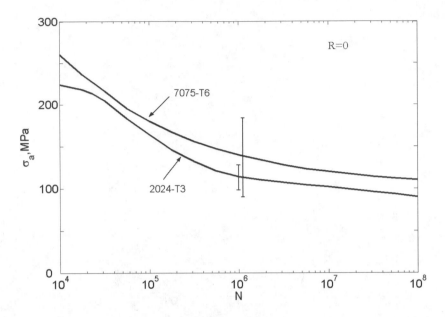

Figure B.3. Fatigue properties of select alloys: (a) S-N curves (R = 0) for 2024-T3 and 7075-T6 representing average of several extensive studies [68–70,96,99]. The brackets show the spread of mean values of two studies (not the dispersion of the measurements) and are an example of the uncertainties introduced by lot-to-lot material variability.

SCF-FS analysis may be useful for shear connections when these are subjected to low, nominal stress amplitudes and display fatigue lives, $N > 10^5 - 10^6$ cycles. These considerations are validated by the work of Seliger [31], who applied the SCF-FL analysis as early as 1943.

Seliger derived SCF-values for riveted lap joints from measurements of the S-N curves of the material and the joint. His findings, recast in the format of Figure B.1, are presented in Figure B.4. Comparison of these two figures reveals that the material/panel S-N curve relations for the lap joint and open hole panels are essentially the same. The main difference is that the reliability of the SCF-FS analysis for this particular lap joint is limited to $N > 2 \cdot 10^6$ compared to $10^5 < N < 10^6$ for open hole panels. As noted above, this is a consequence of the high SCFs of single, fastener-row joints and the limitation on yielding. Additional validation of the SCF-FS analysis for aluminum and steel joints is presented in Chapter 4.3.

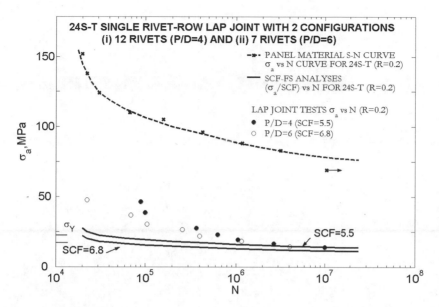

Figure B.4. A comparison of the fatigue test data for 24S-T, single rivet-row lap joints with the predictions of the SCF-FS Analysis [32].

ANALYSIS OF FRETTING WEAR AND FATIGUE

Fretting fatigue is a special case of structural fatigue in which the critical crack initiates at an interface subjected to fretting, and the fatigue life is influenced by the contact conditions, such as the contact pressure and the tangential tractions. Fretting fatigue is suspected when signs of fretting wear are observed near the site of crack initiation. But this is not sufficient evidence for several reasons: (i) fretting can produce wear particles without initiating fatigue cracks, (ii) fretting wear can "erase" initiating cracks, (iii) fretting wear increases after a crack has been initiated due to increased local compliance, and (iv) fretting wear generally does not influence the crack growth stage.

Current understanding of the role of fretting debris in modifying the contact stress field, SCF and fatigue life contact remains limited. Fretting fatigue combines three behaviors which are difficult to analyze or study experimentally: (i) contact mechanics, (ii) wear, and (iii) fatigue crack initiation. Owing to this confluence of three challenging problems, experimental and analytical research on this subject has been concentrated on planar receding contact configurations, i.e., cylinder-on-flat or sphere-on-flat or punch-on-flat, subjected to constant normal loading. Very recent studies with cylinder-on-flat contact configurations [100] indicate that observed reductions in fatigue life due to fretting under elastic contact conditions are related simply to the peak contact pressure, p_0, nominal cyclic stress range, $\Delta\sigma$, and maximum local cyclic stress range, $\Delta\sigma_{L,max}$ as follows:

$$N_i^{FF} \approx N_i^P \cdot \left[\frac{\Delta\sigma_N}{\Delta\sigma_{L,max}} \right]^{a+d} \cdot e^{mp_0} \qquad (C1)$$

Equation (C1) shows that fretting contact reduces plain fatigue life by a Stress Range Concentration Factor (SRCF), defined as SRCF $= \left[\dfrac{\Delta\sigma_{L,max}}{\Delta\sigma_N} \right]$, raised to the power $-(a+d)$, where $-a$ is the fretting fatigue-strength exponent can be determined for any structural contact. It also shows that the peak contact pressure extends the plain fatigue through an exponential relationship and another constant, m. Fretting wear parameters such as slip amplitude and shear traction do not enter Equation (C1) directly because they are related to the peak contact pressure and local cyclic stress range and are therefore already accounted for. The implication is that the contact stress field alone causes the reduction in fatigue life and fretting wear does not have any separate influence. It is significant that this finding is independently substantiated by the SCF-FS approach (Appendix B and section 4.3), which demonstrates a correlation between joint fretting fatigue life measurements and the contact-induced peak stress (SCF) without resort to any wear-related parameters.

To the knowledge of the authors, there are no direct *analyses* of fretting for the contact configurations and loading conditions present in structural shear joints other than those presented in this book. Equation (C1) cannot be used directly unless the constants are determined for the appropriate contact and material pairs. However, experimental and analytical studies of other types of contact (see above) have been performed since the early 1980's to address concerns with aging military and civilian aircraft engines. A pair of parameters developed as design aids for the turbine engine dovetail joint fretting problem, and applied to the structural shear joints presented in this book, are described next.

Assessing the Severity of Fretting Wear and Fretting Fatigue. In the absence of more reliable means of quantitatively assessing the severity of fretting damage, two empirical parameters, F_1 and F_2, which have been found to correlate with fretting wear [31] and fretting fatigue [30] in turbine engine dovetail joints, have been evaluated at the interfaces of the shear joints:

$$\text{Fretting wear factor} \qquad F_1 = \beta \, \mu \, p \, \delta \qquad\qquad\qquad \text{(C2)}$$
$$\text{Fretting fatigue factor} \qquad F_2 = \beta \, \mu \, p \, \delta \, \sigma_{\theta\theta,max} \qquad\qquad \text{(C3)}$$

The quantities p, δ and $\sigma_{\theta\theta,max}$ are the local contact pressure, slip and tangential stress amplitude; the coefficient, β, is introduced here to nominally distinguish between different fretting conditions[1]. The depth of the wear scar, y, is directly related to W_S, the specific wear rate of the material and F_1:

$$y = 2N \, F_1 \, (W_S / \mu) \qquad\qquad\qquad\qquad \text{(C4)}$$

where N is the number of fretting cycles. The importance of fretting fatigue is assessed by comparing the peak value of F_2, generated in the shear joint, with threshold values determined experimentally for fatigue crack initiation. These are $F_2 = 4.2 \times 10^9$ Pa^2m in an Al-4%Cu alloy is [101] and $F_2 > 50 \times 10^9$ Pa^2m [102] in hardened steel.

A number of disclaimers with regard to these two fretting parameters are noteworthy:

• The expression for F_2 does not account for the pressure gradient, which is useful when fretting conditions with comparable gradients are compared.

• The above threshold values apply to the fretting conditions produced by a pressure gradient that is different from those present in structural shear joint interfaces. They are expected to understate the threshold for bearing mode fretting at the periphery of the fastener hole interface locations A and A', where the attending pressure gradients are shallow by comparison.

---

[1] $\beta = 1$: Fretting of nonconforming surfaces under constant p

$\beta = 1/2$ Fretting of nonconforming surfaces for p varying linearly with applied stress

$\beta = 1/4$ Fretting of conforming surfaces for p varying linearly with applied stress. This accounts for the fact that fretting wear diminishes the local contact pressure of conforming surfaces

- F_2 does not have any known physical significance. Its definition is purely empirical.
- Similar values of F_1 and F_2 can be obtained for conditions that show different fretting wear and fatigue life (need to add ref). For example, increasing the normal load leads to a reduction in fatigue life. But because the slip amplitude reduces and the bulk stress increases, the values of the parameters can remain constant, erroneously suggesting no change in structural life.

SYMBOLS, NOMENCLATURE, ABREVIATIONS, AND UNITS

Symbol[1]	Uses	SI Units
A	Projected area of shank, $A = D_St$	mm^2
A_C	Effective area of contact under bolt head	mm^2
A_S	Cross section area of shank	mm^2
C	Excess compliance; $C = C' - C''$	m/GN
C'	Assembly compliance; ratio of net joint to the load per repeat distance, P_1	m/GN
C''	Compliance of a continuous panel having the same length as the joint; length of the 2 (or 3) panels minus the overlap	m/GN
%C	Compression of shank expressed as a percent: $\%C = \%X(1-\mu)/(1+\mu)$	
%CL	Nominal clamping expressed as a percent	
D	Diameter of fastener hole	mm
D_0	Initial diameter of fastener hole	mm
D_B	Diameter of fastener head	mm
D_S	Diameter of fastener shank	
2D, 3D	2-dimensional, 3-dimensional	
E	Elastic tension modulus	GPa
EIP	Elastic-isotropic-plastic constitutive relations	
ELKP	Elastic-linear-kinematic-plastic constitutive relations	
f_i	Fraction of the total fastener load, Q, supported by individual fasteners in the ith–row of a multiple row joint	
f_{BM}	Fraction of the applied stress supported by the bearing mode when $\sigma = \sigma_C$	
f_{BPi}	Fraction of the total fastener load that bypasses the ith row	
F1, F2, F3	Shear transfer loads	N
F_1	Fretting wear parameter	Pam
F_2	Fretting fatigue parameter	Pa^2m
F_{33}	Net axial force supported at the fastener midsection	N
$F_{33,a}$	Axial force amplitude supported at the fastener midsection	N
FEA	Finite element analysis	

[1] See Appendix A for joint abbreviations.

Symbol	Uses	SI Units
FS	Fatigue strength; stress amplitude for fatigue failure after a given number of stress cycles	MPa
h	Thickness of brazed layer	in
H	Length of shank between rivet heads or between bolt head and retaining nut; $H = 2t$	mm
H_0	Length of shank between rivet heads before assembly of a clamped joint; $H_0 < 2t_0$	mm
ΔH	Extension of fastener shank, $\Delta H = \varepsilon\, H_0$	mm
ΔH_L	Local compression of fastener head and the fastener hole edge	mm
%I	Interference expressed as a percent; $\%I = 100(D_S - D_0)/D_0$	
k	Stiffness of fastener/fastener hole response of the 2-springsmodel of a multi-row joint; $k = C^{-1}$	MN/m
K	Stiffness response of intervening panel strip between adjacent rows of fasteners of the 2-springs model of a multi-row joint	MN/m
L	Overall length of joint	mm
L_1, L_2	Overall width and length of panel, respectively	mm
M	Plastic modulus	GPa
MSD	Multiple Site Damage	
	Number of stress cycles; also, number of stress cycles to failure	
N	Number of rivet rows	
N_i, N_G	Number of stress cycles consumed by the crack initiation and growth stages	
N1, N2, N3	normal transfer loads	N
NDI, NDE	Nondestructive Inspection/Evaluation	
NPC	Negative ratio of the peak compressive stress to the nominal tensile stress	
p	Contact pressure	MPa
P_1	Fastener pitch or repeat distance along the row	mm
P_2	Spacing of fastener rows	mm
$p\varepsilon_{mag}$	Equivalent plastic strain	
$p\varepsilon_{\theta\theta}$	Tangential plastic strain	
Q	Total fastener load; load supported by individual fasteners of a single-fastener row joint; total load supported by vertical line of fasteners of a multi-row joint	N
$Q_1, Q_2, Q_i, \ldots Q_n$	Loads transmitted to fasteners in the first, second, i^{th} and n^{th} rows of a multiple row joint	N
Q_a	Fastener load amplitude	N
Q_B	Load transmitted to underside of the fastener head in a lap joint	N
$Q_{BP,A}, Q_{BP,B}, \ldots$	Bypass load for the first, second (and subsequent rows) of fastners	N

Symbol	Uses	SI Units
Q_C	Maximum load a clamped fastener can support by friction	N
Q_i	Load supported by individual fasteners in the ith-row	N
r	Radial distance measured from the center of the fastener hole	mm
r_F	Radius of the head-shank fillet	mm
r_H	Radius of fastener hole	mm
$r_{H,0}$	Initial radius of fastener hole	mm
r_Y	Radius of yielded (plastic) zone	mm
R	Stress ratio; $R = \sigma_{MIN}/\sigma_{MAX}$	
SAF	SCF * N	
SC	Stress concentration	
SCF	Stress concentration factor, $SCF = \sigma^*/\sigma$	
SCF*	SCF for the elevation of the stress range, $SCF^* = (\sigma^* - \sigma_{RESIDUAL})/\sigma$	
SCF_B	SCF of the bearing mode	
$SCF_{BP,i}$	SCF produced by the by-pass load at row i	
$SCF_{FL,i}$	SCF produced by the fastener load at row i	
$SCF_1, SCF_2 \ldots$	SCF produced at 1st. 2nd, and subsequent rows	
SCF-FS	Stress Concentration Factor-Fatigue Strength Analysis	
S-N	Stress amplitude-number of cycles to failure relation	
SRCF	Stress range concentration factor	
SRF	Strength reduction factor; SRF = FS(material)/FS(joint)	
t	Panel thickness	mm
t_0	Panel thickness before assembly	mm
Δt	Panel compression	mm
TALA	Thin Adhesive Layer Analysis	
W_S	Specific wear rate	m^3/Nm
%X	Hole expansion before any load release expressed as a percent; $\%X = 100\,(D - D_0)/D_0$	
x	Coordinate in the plane of the panel and normal to joint (or panel) loading direction	
y	Coordinate in the plane of the panel and in the joint (or panel) loading direction; also, depth of wear scar	mm
z	Coordinate normal to the plane of the panel; panel depth coordinate	
α	Fastener tilt angle	degrees
β	Coefficient introduced to account for different fretting conditions	
δ	micro-slip; relative displacement of contacting surfaces	μm
ε	Average elastic strain of fastener shank after assembly	
ϕ	Stress biaxiality; $\phi = \sigma$ (transverse to loading direction)/σ	
η	Dimensionless factor accounting for the effects of the none uniform stress distribution on K.	
φ	Dimensionless factor relating Q and Q_C, or σ_C and Q_C; $\varphi = 1$ for lap joints, $\varphi = 2$ for butt joints	
μ	Coefficient of friction;	

Symbol	Uses	SI Units
ν	Poisson's ratio	
θ	Angular location measured counter clockwise from 3 o'clock	degrees
σ	Gross section nominal stress	MPa
$\Delta\sigma$	Stress range; $\Delta\sigma = (\sigma_{MAX} - \sigma_{MIN})$	MPa
$\Delta\sigma_{\theta\theta}$	Tangential stress range	MPa
$\sigma_1, \sigma_2, \sigma_3$	Principle stresses, with the 1-, 2- and 3-direction corresponding to the in-plane loading, in-plane transverse and out-of-plane directions	MPa
$\sigma_{11}, \sigma_{22}, \sigma_{33}$	Normal stresses, with the 1-, 2- and 3-directions, corresponding to the in-plane loading, in-plane transverse and out-of-plane directions	MPa
$\sigma_{12}, \sigma_{23}, \sigma_{13}$	Shear stresses	MPa
σ^{*}	Peak σ_{11} tensile stress of the stress concentration	MPa
σ_a	Gross section stress amplitude; $\sigma_a = (\sigma_{MAX} - \sigma_{MIN})/2$	MPa
σ_C	Maximum applied (gross section) stress supported by the clamping mode for a particular value of Q_C	MPa
σ_{CM}	Applied (gross section) stress that is accompanied by a change in the slope of the $\sigma^{*}_N - \sigma$ curve marking the transition from the clamping mode to the freely slipping, clamping-plus-bearing mode. Note: $\sigma_{CM} > \sigma_C$	MPa
σ_m	Mean stress; $\sigma_m = (\sigma_{MAX} - \sigma_{MIN})/2 = \sigma_a (1 + R)/(1 - R)$	MPa
σ_Y	Yield stress	MPa
σ_r	Radial stress	MPa
$\sigma_{\theta\theta}$	Tangential stress	MPa
$(\sigma_{\theta\theta})_m$	Mean tangential stress	MPa
σ_a^{*}	Peak (local) stress amplitude	MPa
$\sigma^{*}_{a,33}$	Peak axial stress amplitude applied to fastener	MPa
σ^{*}_G	Peak σ_{11} tensile stress within the gross section stress concentration	MPa
σ^{*}_m	Peak mean stress	
$\sigma_{MIN}, \sigma_{MAX}$	Minimum and maximum stress of stress cycle	MPa
σ^{*}_N	Peak σ_{11} tensile stress within the net section stress concentration	MPa
$\sigma^{*}_{RESIDUAL}$	Local residual stress produced by clamping	MPa
σ_S	Average, axial tensile stress	MPa
$\sigma_{S,33}$	Average, axial tensile stress acting at the fastener midsection	MPa
ω	$\omega = \exp[(p/\delta_Y) - 0.5]$	
ξ	Dimensionless factor relating the SCF to the SCF_B; $\xi \approx 1.1$	
ψ	Dimensionless factor relating the SCF to the P_1/D-ratio	

REFERENCES

1. Iyer, K., 1997, "Three-Dimensional Finite Element Analyses of the Local Mechanical Behavior of Riveted Lap Joints," Ph.D. Dissertation, Vanderbilt University, Nashville, TN.

2. Dechwayukul, C., 1998, "Load Transfer in Single Row Riveted Lap Joints," M.S. Thesis, Vanderbilt University, Nashville, TN.

3. Loha, C., 2000, "Load Transfer in Steel and Aluminum Riveted Lap Joints," M.S. Thesis, Vanderbilt University, Nashville, TN.

4. Al-Dakkan, K., 2000, "The Effect of Interference on Fatigue Life in Single Rivet-Row, Aluminum Lap Joints," M.S. Thesis, Vanderbilt University, Nashville, TN.

5. Huang, Y. H., 2001, "The Effect of Interference on Fatigue Life in Single and Double Rivet-Row, Aluminum Lap Joints," M.S. Thesis. Vanderbilt University, Nashville, TN.

6. Dechwayukul, C., 2001, "Development of a Thin Adhesive Layer Analysis for Riveted and other Structural Joints," Ph.D. Dissertation, Vanderbilt University, Nashville, TN.

7. Fongsamootr, T., 2001, "The Dilational and Compressive Properties of a Polymer Sealant and Analyses of the Distortion and Fatigue of Sealed Riveted Lap Joints," Ph.D. Dissertation, Vanderbilt University, Nashville, TN.

8. Kamnerdtong, N., 2001, "The Shear Properties of a Polymer Sealant and Analyses of the Distortion and Fatigue of Sealed Countersunk Riveted Lap Joints," Ph.D. Dissertation, Vanderbilt University, Nashville, TN.

9. Iyer, K., Hahn, G. T., Bastias, P. C. and Rubin, C. A., 1995, "Analysis of Fretting Conditions in Pinned Connections," Wear, **181-183**, pp. 524-530.

10. Xue, M., Iyer, K., Kasinadhuni, R., Bastias, P. C., Hahn, G. T., Wert, J. J. and Rubin C. A., 1995, "Fretting of Riveted Aluminum Joints," *Proc. Jordanas SAM95*, Cordoba, Argentina, pp. 495-500.

11. Iyer, K., Xue, M., Kasinadhuni, R., Bastias, P. C., Rubin, C. A., Wert, J. J. and Hahn, G. T., 1995, "Contribution of Fretting to the Fatigue and Corrosive Deterioration of a Riveted Lap Joint," *Structural Integrity in Aging Aircraft*, ASME **AD – Vol. 47**, pp. 35-62. Proc. ASME/IMECE 1995, San Francisco, CA.

12. Iyer, K., Xue, M., Bastias, P. C., Rubin, C. A. and Hahn, G. T., 1996, "Analysis of Fretting and Fretting Corrosion in Airframe Riveted Connections," *Proc. 82nd Meeting of the AGARD SMP on Tribology for Aerospace Systems*, Sesimbra, Portugal, pp. 13/1-13/12.

13. Iyer, K., Bastias, P. C., Rubin, C. A. and Hahn, G. T., 1997, "Analysis of Fatigue and Fretting of Three-Dimensional, Single and Double Rivet-Row Lap Joints," *Proc. ICAF97 Symposium on Fatigue in New and Ageing Aircraft*, **II**, Edinburgh, Scotland, pp. 855-869.

14. Hahn, G. T., Iyer, K., Rubin, C. A. and Bastias, P. C., 1997, "Three-Dimensional Analyses of Fatigue and Fretting Conditions in Aluminum Alloy Riveted Lap Joints," *Proc. First Joint DoD/FAA/NASA Conference on Aging Aircraft,* Ogden, UT.

15. Iyer, K., Rubin, C. A. and Hahn, G. T., 1999, "Three-Dimensional Analyses of Single Rivet-Row Lap Joints - Part I: Elastic Response," *Recent Advances in Solids and Structures*, ASME **PVP - Vol. 398**, pp. 23-39. Proc. ASME/IMECE 1999, Nashville, TN.

16. Iyer, K., Rubin, C. A. and Hahn, G. T., 1999, "Three-Dimensional Analyses of Single Rivet-Row Lap Joints - Part II: Elastic-Plastic Response," *Recent Advances in Solids and Structures*, ASME **PVP - Vol. 398**, pp. 41-57. Proc. ASME/IMECE 1999, Nashville, TN.

17. Iyer, K., Rubin, C. A. and Hahn, G. T., 2001, "Influence of Interference and Clamping on Fretting Fatigue in Single Rivet-row Lap Joints," ASME J. Tribol., **123**, pp. 686-698.

18. Iyer, K., 2001, "Solutions for Contact in Pinned Connections," Int. J. Sol. Struct., **38** (50-51), pp. 9133-9148.

19. Iyer, K., Rubin, C. A. and Hahn, G. T., 2001, "Three-Dimensional Analysis of Double Rivet-Row Lap Joints. Part I: Noncountersunk Rivets," *Reliability, Stress Analysis and Failure Prevention*, ASME **DE - Vol. 114**, pp. 19-31. Proc. ASME/IMECE 2001, New York, NY.

20. Iyer, K., Rubin, C. A. and Hahn, G. T., 2001, "Three-Dimensional Analysis of Double Rivet-Row Lap Joints. Part II: Countersunk Rivets," *Reliability, Stress Analysis and Failure Prevention*, ASME **DE - Vol. 114**, pp. 33-47. Proc. ASME/IMECE 2001, New York, NY.

21. Dechwayukul, C., Rubin, C. A. and Hahn, G. T., 2003, "Analysis of the Effects of Thin Sealant Layers in Aircraft Structural Joints," AIAA Journal, **41**(11), pp. 2216-2228.

22. Iyer, K., Rubin, C. A. and Hahn, G. T., 2004, "Clamping and Failure Mode Transitions in Structural Shear Joints," *Analysis of Bolted Joints*, ASME **PVP - 478**, pp. 1-9. Proc. ASME/JSME Pressure Vessels and Piping Division Conference, San Diego, CA.

23. Hoggard, A. W., 1991, "Fuselage Longitudinal Splice Design," *Structural Integrity of Aging Airplanes*, S.N. Atluri et al., eds., Springer Verlag, Berlin, Germany, pp. 167-183.

24. Swift, T., 1991, "Repairs to Damage Tolerant Aircraft," *Structural Integrity of Aging Airplanes*, S.N. Atluri et al., eds., Springer Verlag, Berlin, Germany, pp. 433-484.

25. Ciavarella, M. and Decuzzi, P., 2001, "The State of Stress Induced by the Plane Frictionless Cylindrical Contact. I. The Case of Elastic Similarity," Int. J. Sol. Struct., **38**, (26-27), pp. 4507-4523.

26. Hahn, G. T., Bhargava, V. and Chen, Q., 1990, "The Cyclic Stress-Strain Properties, Hysteresis Loop Shape, and Kinematic Gardening of Two High Strength Bearing Steels," Met. Trans. A, **21A**, pp. 653-665.

27. McDowell, D. L., 1985, "A Two Surface Theory for Non-Proportional Cyclic Plasticity, Part1: Development of Appropriate Equations," ASME J. Appl. Mech., **52**, pp. 298-302.

28. McDowell, D. L., 1985, "A Two Surface Theory for Non-Proportional Cyclic Plasticity, Part2: Comparison of Theory with Experiments," ASME J. Appl. Mech., **52**, pp. 303-308.

29. Timoshenko, S., 1956, *Strength of Materials Part II: Advanced Theory and Problems*, 3rd edition, Van Nostrand, Princeton, p. 301.

30. Waterhouse, R. B., 1992, "Fretting Wear," *ASM Metals Handbook*, No. 18, pp. 242-256.

31. Ruiz, C., Boddington, P. H. B. and Chen, K. C., 1984, "An Investigation of Fatigue and Fretting in a Dovetail Joint," Exp. Mech., **24**, pp. 208-217.

32. Seliger, V., 1943, "Effect of Rivet Pitch Upon the Fatigue Strength of Single-Row Riveted Joints of 0.025- to 0.025-inch 24S-T ALCLAD," Technical Report No. 900, National Advisory Committee for Aeronautics, Washington, D. C.

33. Fongsamootr, T. (unpublished results).

34. Muller, R. P. G., 1995, "An Experimental and Analytical Investigation on the Fatigue Behavior of Fuselage Riveted Lap Joints: The significance of the Rivet Squeeze Force and a Comparison of 2024-T3 and Glare3," Ph.D. Dissertation, Technical University of Delft, The Netherlands.

35. Iyer, K. (unpublished results).

36. Timoshenko, S., 1955, *Strength of Materials Part I: Elementary Theory and Problems*, 3rd edition, Van Nostrand, Princeton, p. 112.

37. Sharp, M. L., Nordmark, G. E. and Menzemer, C. C., 1996, *Fatigue Design of Aluminum Components and Structures*, McGraw Hill, New York.

38. Mettu, S. R., DeKoning, A. U., Lof, C. J., Schra, L., McMahon, J. J. and Forman, R. G., 2000, "Stress Intensity Factor Solutions for Fasteners in NASGRO 3.0," *Structural Integrity of Fasteners: Second Volume*, ASTM STP 1391, P. M. Toor, ed., ISBN 0-8031-2863-0.

39. Kephart, A. R., 2000, "Fatigue Acceptance Test Limit Criterion for Larger-Diameter Rolled Test Fasteners," *Structural Integrity of Fasteners: Second Volume*, ASTM STP 1391, P. M. Toor, ed., ISBN 0-8031-2863-0.

40. Bickford, J. H., 1995, *An Introduction to the Design and Behavior of Bolted Joints*, 3rd edition, Marcel Dekker, Inc., New York, pp. 151-158.

41. Chesson, E. Jr., and Munse, W. H., 1965, "Studies of the Behavior of High Strength Bolts," Eng. Exp. Bull., University of Illinois at Urbana-Champaign.

42. Sterling, G. H., Troup, E. W. J., Chesson, E. Jr., and Fisher, J. W., 1965, "Calibration Tests of A490 High Strength Bolts," ASCE J. Struct. Div., **91** (ST5), pp. 279-287.

43. Christopher, A. M., Kulak, G. L., and Fisher, J. W., 1966, "Calibration of Alloy Steel Bolts," ASCE J. Struct. Div., **92** (ST2), pp. 19-41.

44. Iyer, K., Brittman, F. L., Hu, S. J., Wang, P. C., Hayden, D. B. and Marin, S. P., 2002, "Fatigue and Fretting of Self-Piercing Riveted Joints," ASME **MED - Vol. 13**, pp. 401-415. Proc. ASME/IMECE 2002, New Orleans, LA.

45. Ozelton, M. N. and Coyle, T. G., 1986,"Fatigue Life Improvement by Cold Working Fastener Holes in 7050 Aluminum," *ASTM STP 927*, J.M. Potter, ed., ASTM, Philadelphia, pp. 53-71.

46. Szolwinski, M. P. and Farris, T. N., 2000, "Linking Riveting Process Parameters to the Fatigue Performance of Riveted Aircraft Structures," J. Aircraft, **37**, pp. 130-137.

47. Szolwinski, M.P., Harish, G., McVeigh, P.A. and Farris, T. N., 2000, "Experimental Study of Fretting Crack Nucleation in Aerospace Alloys with Emphasis on Life Prediction," *ASTM STP 1367*, D.W. Hoeppner et al., eds., ASTM, West Conshohocken, pp. 267-281.

48. Timoshenko, S., 1956, *Strength of Materials Part II: Advanced Theory and Problems*, 3rd edition, Van Nostrand, Princeton, pp. 210-213.

49. Park, J. H. and Atluri, S. N., 1993, "Fatigue Growth of Multiple-Cracks Near a Row of Fastener Holes in a Fuselage Lap Joint," Comp. Mech., **13**, pp. 159-203.

50. Deng, X. and Hutchinson, J. W., 1998, "The Clamping Stress in a Cold-Driven Rivet," Int. J. Mech. Sci., **40**, pp. 683-694.

51. Kulak, G. L., Fisher, J. W. and Struik, J. H., 1987, *Guide to Design Criteria for Bolted and Riveted Joints, Second Edition*, John Wiley and Sons, New York, pp. 116-125.

52. Schijve, J., 1992, "Multiple Site Damage of Riveted Joints," *Durability of Metal Aircraft Structrures*, S. N. Atluri et al., eds., Atlanta Technology Publications, Atlanta, GA, pp.2-27.

53. Birkemoe, B. C., Meinheit, D. F. and Munse, W. H., 1969, "Fatigue of a 514 Steel in Bolted Connections," ASCE J. Struct. Div., **95**, pp. 2011-2030.

54. Birkemoe, B. C. and Srinivasan, R., 1971, "Fatigue of Bolted High Strength Structural Steels," ASCE J. Struct. Div., **97**, pp. 935-950.

55. Fawaz, S. A., 1997, "Fatigue Crack Growth in Riveted Joints," Ph.D. Dissertation, Technical University of Delft, The Netherlands.

56. Okada, T., 2001, "Fatigue Behavior of Lap Joint Fuselage Model Structure," *Proc. 5th Joint FAA/DOD/NASA Conference of Aging Aircraft*, Orlando, FL.

57. Wang, L., Brust, F. W. and Atluri, S. N., 1997, "Predictions of Stable Growth of Lead Crack and Multiple-Site Damage Using Elastic-Plastic Finite Element Alternating Method (EPFEAM)," *Proc. FAA-NASA Symp. On Continued Airworthiness of Aircraft Structures*, **2**, Report DOT/FAA/AR-97/2, I., C. A. Bigelow, ed., pp. 505-518.

58. Mar, J. W., 1991, "Structural Integrity of Aging Airplanes: A Perspective," *Structural Integrity of Aging Airplanes*, S. N. Atluri et al., eds., Springer Verlag, Berlin, Germany, pp.241-262.

59. Pyo, C. R., Okada, H., Wang, L., Brust, F. W. and Atluri, S. N., 1995, "Residual Strength Prediction For Aircraft Panels With Multiple Site Damage (MSD) Using the Elastic-Plastic Finite Element Alternating Method (EPFEAM)," *Structural Integrity in Aging Aircraft*, ASME **AD - Vol. 47**, pp. 73-80.

60. Silva, L. F. M., Goncalves, J. P. M., Oliveira, F. M. F. and de Castro, P. M. S. T., 2000, "Multiple-Site Damage in Riveted Lap-Joints; Experimental Simulation and Finite Element Prediction," Int. J. Fatigue, **22**, pp. 319-338.

61. Ingraffea, A. R., Grigoriu, M. D. and Swenson, D. V., 1991, "Representation and Probability Issues in the Simulation of Multi-Site Damage," *Structural Integrity of Aging Airplanes,* S. N. Atluri et al., eds., Springer Verlag, Berlin, Germany, pp.183-198.

62. Pártl, O. and Schijve, J., 1993, "Multiple-Site Damage in 2024-T3 Alloy Sheet," Int. J. Fatigue, **15** (4), pp. 293-299.

63. Harris, C. E., Newman, J. C. Jr., Piascik, R. S. and Starnes, J. H. Jr., 1997, "Analytical Methodology for Predicting the Onset of Widespread Fatigue Damage in Fuselage Structure," *Proc. FAA-NASA Symp. On Continued Airworthiness of Aircraft Structures,* **1**, Report DOT/FAA/AR-97/2, I., C. A. Bigelow, ed. pp. 63-87.

64. Wang, L., Chow, W. T., Kawai, H. and Atluri, S. N. 1997, "Predictions of Widespread Fatigue Damage Threshold," *Proc. FAA-NASA Symp. On Continued Airworthiness of Aircraft Structures*, **2**, Report DOT/FAA/AR-97/2, I. C. A. Bigelow, ed., pp. 519-530.

65. *SAE Fatigue Design Handbook,* J. A. Graham, ed., SAE, New York, p. 55.

66. Frost, N. E., Marsh, K. J. and Pook, L. P., 1974, *Metal Fatigue*, Clarendon Press, Oxford, pp.176-178.

67. Hookham, C. R., 1979, "Endurance of Riveted Lap Joints (Aluminum Alloy Sheets and Rivets," Report of the Strength and Analysis Group of EDSU, pp. 9, 10, 12 and 13.

68. MIL-Handbook-5H 1, Dec.1998, p. 3-115.

69. Grover, H. J., Hyler, W. S., Kuhn, P., Laners, C. B. and Howell, F. M., 1953, "Axial Load Fatigue Properties of 24S-T and 75S-T Aluminum Alloy as Determined in Several Laboratories," Technical Report No. NACA TN-2928, National Advisory Committee for Aeronautics, Washington, D. C.

70. *ASM Handbook, Fatigue and Fracture,* (Design for Fatigue Resistance), **19**, p. 519.

71. Haarjer, G., 1966, "Design Data for High Yield Strength Alloy Steel," ASCE J. Struct. Div., **92**, pp. 31-49.

72. Fountain, R. S., Munse, W. H. and Sunbury, R. D., 1968, "Specifications and Design Relations," ASCE J. Struct. Div., **94**, pp. 2751-2765.

73. Mindlin, H., 1968, "Influence of Details on Fatigue Behavior of Structures," ASCE J. Struct. Div., **94**, pp. 2679-2697.

74. Hansenm N. G., 1959, "Fatigue Tests of Joints of High Strength Steels," ASCE J. Struct. Div., **85**, pp. 51-69.

75. Hirt, M. A., Yen, B. T. and Fisher, J. W., 1971, "Fatigue Strength of Rolled and Welded Steel Beams," ASCE J. Struct. Div., **97**, pp. 2011-2028.

76. Chesson, E. Jr. and Munse, W. H., 1965,"Studies of the Behavior of High Strength Bolts and Bolted Joints," Exp. Sta. Bull., University of Illinois at Urbana-Champaign, **469**, pp. 1- 35.

77. Albrecht, P. and Sahil, A. H., 1986, "Fatigue Strength of Bolted and Adhesive Bonded Structural Steel Joints," *Fatigue in Mechanically Fastened Composite and Metal Joints, ASTM STP 927,* J. M. Potter, ed., ASTM, Philadelphia, pp. 72-94.

78. Hookham, C. R., 1984, "Endurance of Aluminum Alloy Lugs with steel Interference-Fit Pins and Bushes," Technical Report No. 84025, Strength Analysis Group of ESDU, pp. 1-31.

79. Morgan, F. G., 1962, "Static Stress Analysis and Fatigue Tests of Interference-Fit Bushes," RAE Tech. Note, Structures, **316**, Farnsborough, U.K.

80. Wallgren, G., 1965, "Utmattningshall-Fasteten Hos Aluminiumoron Bussningar," Tech. Note HU-1068, FFA, Aero. Res. Inst., Sweden.

81. Luo, A. A., Kubic, R. C. and Tartaglia, J. M., 2003, "Microstructure and Fatigue Properties of Hydroformed Aluminum Alloys 6063 and 5754," Metals and Minerals Trans. A (Wilson Appl. Sci. and Tech. Abstracts), **34A**, pp. 2549-2557.

82. Iyer, K., Xue, M., Kasinadhuni, R., Bastias, P.C., Rubin, C.A., Wert, J.J. and Hahn, G. T., 1996, "Contribution of Fretting to the Fatigue and Corrosive Deterioration of a Riveted Lap Joint," *Proc. Air Force 3rd Aging Aircraft Conference*, pp. 191-225.

83. Kissell, J. R. and Ferry, R. L., 2002, *Aluminum Structures: A Guide to Their Specification and Design*, John Wiley & Sons, Inc., New York.

84. *Specification for Structural Joints Using ASTM A325 or A490 Bolts*, 2000, Publication of the Research Council on Structural Connections, American Institute of Steel Construction, Inc., Chicago, IL.

85. *Aluminum Design Manual*, 1994, The Aluminum Association, Inc., Washington, D. C.

86. *ABAQUS Users Manual*, 2000, version 5.8.2, Hibbit, Karlsson & Sorensen, Inc.

87. Carter, J. W., Lenzen, K. H. and Wyly, L. T., 1955, "Fatigue in Riveted and Bolted Single-Lap Joints," Trans. ASCE, **120**, pp. 1353-1388.

88. Fung, C. P. and Smart, J., 1994, "An Experimental and Numerical Analysis of Riveted Single Lap Joints," Proc. Inst. Mech. Eng., Part G, J. Aero. Eng., **208**, pp. 79-90.

89. Fung, C. P. and Smart, J., 1997, "Riveted Single Lap Joints; Part I: A Numerical Parametric Study," Proc. Inst. Mech. Eng., Part G, J. Aero. Eng., **211**, pp. 13-27.

90. Cornell, R. W., 1953, "Determination of Stresses in Cemented Lap Joints," J. Appl. Mech., **75**, pp. 355-364.

91. Baker, M. R. and Hatt, F., 1973, "Analysis of Bonded Joints in Vehicular Structures," AIAA Journal, **11**(12), pp. 1650-1654.

92. Huth, H., 1986, "Influence of Fastener Flexibility in the Prediction of Load Transfer and Fatigue Life of Multiple Row Joints," ASTM STP 927, pp. 221-250.

93. Tate, M. B. and Rosenfield, S. J., 1946, "Preliminary Investigation on Loads Carried by Individual Bolts in Bolted Joints, Technical Report No. NACA TN-1052, National Advisory Committee for Aeronautics, Washington, D. C.

94. Swift, T., 1971, "Development of the Fail-Safe Design Features of the DC-10," ASTM STP 486, ASTM, Philadelphia, pp. 164-214.

95. Wang, H. L., Buhler, K. and Grandt, A. F. Jr, 1995, "Evaluation of Multiple Site Damage in Lap Joint Specimens," *Proc. USAF Structural Integrity Program Conf.*, G. K. Waggoner, J. W. Lincoln and J. L. Rudd, eds., San Antonio, TX.

96. Iyer, K., Bastias, P. C., Rubin, C. A. and Hahn, G. T., 1996, "Local Stresses and Distortions of a Three-dimensional, Riveted Lap Joint," *Proc. 1995 USAF Structural Integrity Program Conference*, G.K. Waggoner et al., eds., WL-TR-96-4093, pp. 165-185. Also in Technical Report AF TR-96-4093, pp. 21-38.

97. Gerber, H., 1874, "Bestimmung der zulassigen Spannungen in Eisen- konstuctionen." Z. Bayerischen Architeckten und Ingenieur-Vereins, **6**, pp. 101-110.

98. Forrest, P.G., 1962, *Fatigue of Metals*, Pergamon Press, London, pp.100- 101.

99. Newman, J. C. Jr., Phillips, E. P. and Swain, M. H., 1997, "Fatigue Life Prediction Methodology Using Small Crack Theory and Crack Closure Model," *Proc. FAA-NASA Symposium on the Continued Airworthiness of Aircraft Structures*, **2**, pp. 331-356.

100. Iyer, K., 2001, "Peak Contact Pressure, Cyclic Stress Amplitude, Contact Semi-Width and Slip Amplitude: Relative Effects on Fretting Fatigue Life," Int. J. Fatigue, **23**, pp. 193-206.

101. Nowell, D. and Hills, D. A., 1990, "Crack Initiation Criteria in Fretting Fatigue," Wear, **136**, pp. 329-343.

102. Kuno, M., Waterhouse, R. B., Nowell, D. and Hills, D. A., 1989, "Initiation and Growth of Fretting Fatigue Cracks in the Partial Slip Regime," Fatigue and Fract. Eng. Mat. Struct., **12** , pp. 387-398.

INDEX